Melanie Lott
Biomechanics of Dance

Also of Interest

Klassische Mechanik
Vom Weitsprung zum Marsflug
4. Auflage
Müller, 2021
ISBN 978-3-11-073538-3, e-ISBN (PDF) 978-3-11-073078-4,
e-ISBN (EPUB) 978-3-11-073084-5

Dynamical Systems and Geometric Mechanics
An Introduction
2nd Edition
Maruskin, 2018
ISBN 978-3-11-059729-5, e-ISBN (PDF) 978-3-11-059780-6,
e-ISBN (EPUB) 978-3-11-059803-2

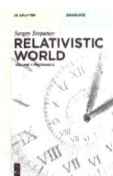

Relativistic World Volume 1: Mechanics
Stepanov, 2018
ISBN 978-3-11-051587-9, e-ISBN (PDF) 978-3-11-051588-6,
e-ISBN (EPUB) 978-3-11-051600-5

Physik für das Lehramt
Band 1: Mechanik und Wärmelehre
Nienhaus, 2017
ISBN 978-3-11-046912-7, e-ISBN (PDF) 978-3-11-046913-4,
e-ISBN (EPUB) 978-3-11-046917-2

Melanie Lott

Biomechanics of Dance

———

Applications of Classical Mechanics

DE GRUYTER

Author
Dr. Melanie Lott
Denison University
100 West College Street
Granville OH 43023
U.S.A.
lottm@denison.edu

ISBN 978-3-11-064228-5
e-ISBN (PDF) 978-3-11-064229-2
e-ISBN (EPUB) 978-3-11-064233-9

Library of Congress Control Number: 2023938205

Bibliographic information published by the Deutsche Nationalbibliothek
The Deutsche Nationalbibliothek lists this publication in the Deutsche Nationalbibliografie;
detailed bibliographic data are available on the Internet at http://dnb.dnb.de.

© 2023 Walter de Gruyter GmbH, Berlin/Boston
Cover image: Maria Daniela Quintero / iStock / Getty Images
Typesetting: VTeX UAB, Lithuania
Printing and binding: CPI books GmbH, Leck

www.degruyter.com

For Trevor and Cameron.
In memory of Ken Laws, my mentor and physics of dance pioneer.

Contents

1 Dance and biomechanics

Dance is as diverse as the many cultures around the globe, but what connects its countless variations is that all dance is a form of expression through body movement. Like all physical objects, motion of the human body follows the principles of physics, and since dancers are *living* physical objects, physics, combined with anatomical and physiological factors, determines a dancer's motion. Due to its expressive component, however, so much of dance is that which is not physical, like the emotional connection between two ballet dancers and an audience during a passionate pas de deux, the joy shared by people dancing to music at a neighborhood block party, or the celebration of life conveyed by funeral dancers in many cultures.

With the physicality of dance, it can certainly be argued that dancers are athletes, and one sub-style of dance – breaking – was confirmed by the International Olympic Committee to make its Olympic debut in 2024. Biomechanics, a branch of biophysics and the study of how and why living things move, has improved athletic performance and reduced injury risk in sports for decades. Performance enhancement motivating the application of biomechanics in sports is straightforward to understand. Many basic movements within athletics have well-defined, objective goals. The Olympic motto, "Citius, Altius, Fortius" ("Faster, Higher, Stronger"), highlights these types of athletic aims. To win the long jump event in track and field, for example, an athlete must leap the furthest distance without considering the aesthetics of the jump. Long jump technique, like other sports, has evolved to enhance this singular aspect of performance under explicit, objective rules of the event.

The desire to prevent injuries also strongly drives biomechanics applications to sport. Injuries cost athletes and teams time, money, and more, and some injuries can be career-ending. Athletes (including dancers) often push their bodies to the limits for peak performance. Dancers, like all athletes, deserve access to innovative, research-based approaches to training and pedagogy so these limits can be safely approached but not exceeded. Biomechanics research can reveal musculoskeletal stresses on the body and suggest modifications to technique, equipment (e. g., footwear, flooring, etc.), and training to reduce injury risk. With its artistic tradition, science-based research practices have been slower to be adopted into dance than into many other sports. Fortunately, acceptance of and enthusiasm for biomechanics research has been growing within the dance community.

The combined athletic and artistic elements of dance present unique challenges for biomechanical applications. Not often is there a single quantity to be optimized in dance like the distance of a long jump. It could be true, for example, that a dancer wishes to improve a vertical leap by jumping higher. A higher jump can appear more impressive to an audience or may be required choreographically. However, achieving maximum height could negatively impact the dancer's overall performance if it detracts from the leap's visual appeal or diverges from the accepted technique of that dance style. These additional factors place constraints on modifications that can be made in the effort to

https://doi.org/10.1515/9783110642292-001

enhance dance performance. Effective biomechanical analyses must account for these constraints: for example, that the skill must appear effortless, that ballet turnout must be maintained, or that an Irish dancer's arms must remain motionless at their sides. The constraint that aesthetic quality must be preserved cannot be ignored, so biomechanics researchers must have expertise in the given dance form and/or consult with those who do. The artistic, subjective nature of dance should not deter biomechanics applications; in fact, science-based approaches to improve athletic aspects can afford dancers more time to focus on artistic enhancement.

The overall movement vocabulary of many dance styles is incredibly diverse, adding to the challenge for biomechanics of dance. Dance occurs in many planes of motion and often involves complex coordination and timing between segments of the entire body. Additionally, there may not exist one standard way to perform a movement depending on the dance style. Classical ballet, with its strict technical vocabulary, has historically been the subject of most biomechanics of dance research. With little room for interpretation, there is often a kinematically correct or optimal way to perform a ballet movement. Like sports with well-defined goals, however, what underlies the kinematics (magnitudes of the forces producing the motion, which muscles are active and when, how the motion is initiated neurologically, etc.), may not be known and can vary between individuals and trials. In biomechanics it is important to consider what works for one individual may not be ideal for another based on a variety of factors including age, strength, anatomy, training, etc.

While there is still much to learn about ballet, biomechanics research has extended and should continue to extend its reach to a more diverse assortment of dance forms, including contemporary, tap, hip-hop, and Irish dance, to name just a few. Similar base movement categories often exist across multiple forms of dance (e.g., jumps, turns, kicks, inversions, etc.), and the same fundamental biomechanical principles apply, independent of dance style. Similar concerns can also arise including overtraining, tradeoffs between flexibility and strength, and types of injuries for example. At the same time, different styles may emphasize movement of different parts of the body, have unique aesthetic aims and considerations, and may place different physical or physiological stresses on the body. Broadening our knowledge base can reveal connections between various styles or what distinguishes one dance form biomechanically from another and would extend the impact and benefits of biomechanics across the dance community.

Dancing through pain, even to the point of detriment to the dancer, and overuse injuries are common in dance, independent of dance style [33]. Dancing through some discomfort is normal and necessary for improvement. If a dancer stopped every time their shoes gave them a blister, they may never dance! Dancers must be attuned to their bodies and have sufficient access to education and specialists to help them determine when pain is a problem or signaling the development of an overuse injury. In recent decades, dance wellness programs have been increasingly integrated into higher education institutions. Many of these programs include such offerings as dance wellness specialists who work with dance students or course offerings for dancers on topics such as anatomy, kinesiology, dance injuries, or biomechanics [15]. Although progress has been made in recent years to integrate dance science and wellness into overall dance education, work should continue to expand on these types of programs.

Beyond the direct benefits to the dance community, dance biomechanics research may lead to basic scientific knowledge about human movement more generally. Dancers are movement experts capable of impressive feats: from soaring through the air, spinning on one foot for many revolutions, suddenly stopping to balance in a challenging posture, or coordinating nuanced movements of essentially any part of the body. Learning not only how dancers accomplish these highly athletic and complex movement tasks, but how dancers might differently perform basic, everyday tasks compared to non-dancers can prove enlightening. For example, research has shown that dancers exhibit different dynamic patterns of postural sway [89] and use different muscle coordination patterns during normal walking [87] than non-dancers. Other studies have found that dancers take longer to change kinematics and kinetics of jump landings due to fatigue [62] and suffer fewer ACL injuries [61] than athletes in other sports. The use of dance as an intervention has also been shown to improve biomechanical outcomes for individuals with neurological disorders such as cerebral palsy, multiple sclerosis, and Parkinson's disease [88]. One reason for these improvements may be that dance instruction breaks complex movements into parts, then later asks dancers to connect them more smoothly, consistent with the theory of dynamic primitives in motor control.

The opportunity for dance biomechanics applications abounds, and this book is intended especially for those interested in this highly interdisciplinary intersection between science and art. The book broadly covers principles of biomechanics rooted in a physics framework. That is to say that it begins with the most fundamental ideas and more simplified physical models of the human body that later build in complexity as the book progresses. Taking complex questions or problems and simplifying them to the most crucial elements helps us gain an understanding of how, and especially why, the human body moves under different conditions. While special attention is given to dance, the tools and techniques covered in this book can be applied to virtually any physical activity. The book may be used as the text for an upper-level, undergraduate physics elective or for a standard biomechanics course. The content and organization of the material was used in a course cross-listed between Physics and Health, Exercise, & Sports Studies at Denison University (PHYS 245/HESS 390: Biomechanics of Sport). Thank you to my students for providing feedback on the readings, which became sections of this book.

This book assumes readers have some basic background in calculus-based Newtonian mechanics (linear and rotational kinematics and dynamics, momentum, and energy), but no prior knowledge of human anatomy, physiology, or even dance is expected. Basic structure and function of the human musculoskeletal and nervous systems are covered in Chapter 2. Physics principles as well as biological elements contributing to the analyses are woven throughout the remainder of the book. For example, in Chapter 3, both the impulse-momentum theorem and the stretch-shortening cycle (when an activated muscle stretches before it shortens) are used to discuss how a dancer can jump higher in a leap. Example problems and solutions are also provided in each chapter, giving readers the opportunity to practice biomechanics problem solving.

Chapter 3 begins with the simplest physical model of a dancer (a point particle) and covers how and why a dancer's center of mass moves under the influence of forces from outside of the body. The next simplest model (inverted simple pendulum) forms the foundation of our inquiry into one of the most central of all dance tasks: balance (Chapter 4). Later in Chapter 4, we consider multi-segmental models (e. g., double or triple inverted pendulum) and discuss coordination strategies for balance maintenance. We extend our analyses in Chapter 5 to modeling the dancer as a rigid body, or a combination of rigid body segments, studying rotations of the whole body (e. g., pirouettes) and individual body segments around joint axes (e. g., rotating the leg around the hip joint as in grand battement). Chapter 6 shifts our focus away from forces outside of the body to musculoskeletal forces inside of the body. We discuss methods to create physical models using knowledge of human anatomy, how to use these models to estimate muscle-tendon forces and joint forces, as well as the limitations of these techniques. In Chapter 7, we consider work and mechanical energy of the moving human body, and we broaden our energy analysis to include metabolic energy requirements and cardiovascular fitness in dance. Chapter 8 is devoted to the more advanced topic of Lagrangian mechanics for readers interested in diving even deeper into human movement analysis. Where the level of mathematical complexity increases (e. g., in sections of Chapter 5 on the inertia tensor and rotational transformations and in Chapter 8), readers are encouraged to consult a reference such as *Mathematical Methods in the Physical Sciences*, by Mary L. Boas [12].

This book demonstrates how readers can combine their understanding of physics with new knowledge about the body to analyze dance movements. Theoretical analyses are supported with examples from current scientific research and demonstrate how theory can be put into practice to improve dance technique and prevent injuries. Additionally, actual experimental data were collected specifically for various examples in the book using standard biomechanics equipment (e. g., motion capture system, electromyography, force plates). Thank you to my undergraduate research students Hoang Anh Nguyen, Warren Xia, and Andy Zabinski, who assisted by working on many of the theoretical and experimental problems for the book. Basic experimental and data analysis methods are also discussed, however, readers interested in learning more about research methods in biomechanics (e. g., measurement techniques, signal processing, computational methods for large data sets, etc.) are encouraged to consult a supplementary resource such as *Research Methods in Biomechanics*, by Robertson, et al. [83].

My own pursuit of dance biomechanics began when I was introduced to the book *Physics and the Art of Dance*, written by Ken Laws, who then became my mentor. I hope that *Biomechanics of Dance: Applications of Classical Mechanics* allows students of physics to see classical mechanics in a new and exciting light, and that it helps guide and inspire the future generation of dance biomechanists.

2 The human body: Musculoskeletal and nervous systems

The human body, with its multiple biological systems and functions, hundreds of bones and muscles, and vast network of neurons, is an extraordinarily complex machine. Creating a simplified model to represent the body is an essential, yet sometimes daunting, task for a biomechanical analysis. For a model to be effective, its simplifying assumptions must be based on appropriate knowledge of the structure and function of the human body.

The purpose of this chapter is to introduce the reader to basic human anatomy and physiology, focusing on the musculoskeletal and nervous systems. Special attention will be made to issues of specific interest in dance, such as musculoskeletal considerations in dance injuries, hypermobility, and comparative research between dancers and non-dancers. Concepts from this chapter are further integrated with mechanical concepts throughout the book. This chapter is not intended to be an exhaustive reference, but rather an elementary foundation for readers without a background in functional anatomy. Readers are encouraged to consult supplemental human anatomy references for more in-depth information.

In addition to understanding the basic mechanical structure of the human body and fundamentals of motor control, it is important for readers to become versed in anatomical vocabulary necessary to communicate with others in the biomechanics field. Use of a common language between individuals and groups is important when sharing information, such as describing a movement or communicating the results of a study. At the same time, biomechanics is a very interdisciplinary field and includes individuals from a variety of disciplinary backgrounds, so overwhelming jargon should be avoided.

Basic dance movements may be common language in the dance community (e. g., plié or relevé for ballet dancers), but to translate these terms for the biomechanics community, specific anatomical terms are needed. The meaning of the French words plié (bent) and relevé (raised), hint to the meaning in dance; however, those not fluent or conversational in the language of ballet can be left wondering what is bent or raised in these movements? This chapter seeks to provide readers with sufficient basic anatomical vocabulary that they can become conversational in the language of anatomy.

2.1 Basic anatomical terminology

Imagine a ladybug lands on you while you are on the phone with your friend. What luck! Before you count its spots to see how many years of good luck you will have, your friend asks you to describe where the ladybug landed. You say that it is on your arm, but your friend wants you to be more specific. You consider referencing a nearby landmark. Is it close to the wrist? The elbow? Then you have a better idea to take a picture or turn on

https://doi.org/10.1515/9783110642292-002

a video chat. That idea quickly fades, however, with your worry that you will startle the bug, and it will fly away. Is it on the top of your arm? Perhaps "top" means maximum vertical position from the ground, but you quickly realize what part of your arm is most vertical depends on whether you are standing up or laying down as well as how you are holding your arm. Even though the ladybug is at rest relative to your arm, you can easily change the position of the ladybug relative to a coordinate system fixed to the ground by changing the position or orientation of your arm. To remedy this situation, we can assign coordinate axes that are fixed relative to the body while in a specific reference position.

The most common reference position of the body is *anatomical position* (AP), in which a person stands with their feet hip width apart and the arms down to the side with palms facing forward (Fig. 2.1). The origin of the body's coordinate system is located at the center of mass. The *mediolateral axis* (labeled X in Fig. 2.1) is horizontal and runs left-right across the body. The *anteroposterior axis* (Y) is also horizontal, but runs in the back-front direction (orthogonal to X), pointing out from the abdomen. With the body standing upright, the *longitudinal axis* (Z) aligns with the vertical line of gravity.

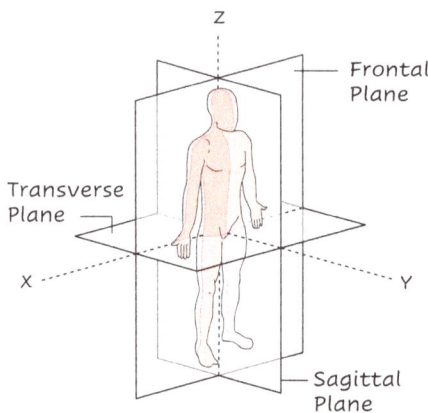

Figure 2.1: A person standing in anatomical position with anatomical axes (X = mediolateral, Y = anteroposterior, Z = longitudinal) and planes shown. "Anatomical planes" by "CFCF" is licensed under CC BY/Labels added.

Certain terms provide descriptors for relative positions of landmarks in these anatomical directions. In the mediolateral (X) direction, *medial* means closer to the longitudinal axis (the midline), and *lateral* means further from the longitudinal axis (a more positive or more negative value of X). For example, your eyes are lateral to your nose. The eyes are medial to your ears, however, because the eyes are closer to the body's midline than the ears. For positions in the anteroposterior (Y) direction, *anterior* means toward the front of the body, and *posterior* means toward the back of the body. For positions in the longitudinal (Z) direction, *superior* means toward the head and *inferior* means toward the feet. Note that since these terms are relative, a landmark or a body part can be on the top half of the body and be termed "inferior" or on the bottom half of the body and termed "superior." For example, the neck is inferior to the head and the knees are superior to the feet. The same is true for the front/back halves of the body and the terms anterior and posterior.

Additionally, there are descriptors specific to the limbs (arms and legs). Positions closer to the body segment's attachment point (e. g., shoulder for the upper arm and hip for the thigh) are termed *proximal*, and positions further from the attachment point are called *distal*. Terms to describe positions relative to the surface of the body are *superficial* (closer to the surface) and *deep* (further from the surface, more internal) and are useful to, for example, describe relative positions of muscles within the body. Many other anatomical terms such as these exist, but may be specific to certain regions of the body (dorsal/plantar for the foot) or used more widely for other organisms (such as cranial/caudal for head/tail) so will not be covered here.

Names are also given to the three planes formed by the anatomical axes: the transverse, sagittal, and frontal planes. The *transverse plane* (XY) is horizontal and divides the body into top and bottom sections. The other two planes are vertical. The *sagittal plane* (YZ) divides the body into left and right halves, and the *frontal plane* (ZX) divides the body into front and back sections. Alternative names for these planes exist (e. g., coronal plane for ZX); however, this book will use the terms transverse, sagittal, and frontal for consistency.

Like most animals, humans exhibit *bilateral symmetry*. This means that reflection of the body across the sagittal plane forms a nearly identical image. There are terms to describe whether two positions are on the same or opposite sides of this symmetry plane. *Ipsilateral* refers to positions on the same side of the body, whereas *contralateral* means on the opposite side of the body. A simple example is that the left arm and left leg are ipsilateral, but the left arm and right leg are contralateral.

As previously described, anatomical axes and planes are useful when describing landmarks on an unmoving body. They can additionally be useful to describe the body in motion. While many dance movements are multiplanar, some occur primarily in a single plane. Motion in one anatomical plane occurs when the body or parts of the body rotate around the axis orthogonal to that plane. For example, rotation round the mediolateral axis (X) results in sagittal plane motion. If a dancer were told they had to limit their movements to motion in the sagittal plane, what could they do? They could do things like bend forward or backward at the waist, kick the leg to the front or back, or walk forward or backward, to name a few. Frontal plane motion occurs when rotation is around the anteroposterior axis (e. g., cartwheel) and motion in the transverse plane occurs when rotation is around the longitudinal axis (e. g., a spin or pirouette). Rotations of parts of the body around axes parallel to X, Y, or Z lead us to a discussion of joint actions.

2.2 Joint actions

Now that we have established some standard anatomical terminology to describe *positions* of landmarks on the body, we move on some common anatomical terms to describe the *motion* of body segments. Dancing can consist of isolated movements of a single body segment, but more often involves numerous body segments in multi-joint motion. Even if the dance movement is complex, it can be very useful to have a standardized way to communicate the primary motions happening around each joint. For the purposes of this section, we can think of a joint as the point of connection between adjacent body segments and the origin for their axes of rotation. (Note this conceptualization of a joint is oversimplistic, and joints will be discussed in greater detail in later sections.)

Joint action is a term to describe the direction of relative motion between two body segments connected at a joint. For each joint action described, we will assume the body begins in anatomical position. We will also imagine a coordinate system with its origin at the joint of interest and axes that are parallel to the body's *XYZ* coordinate system. Terms for joint actions are characterized by rotation about one of these joint axes. In general, 3D rotations of segments around joints are much more complex than described in this section. To quantify rigid body rotations requires well-defined coordinate systems, careful definitions of joint angles, and a 3×3 rotation matrix (discussed in Chapter 5).

2.2.1 Segmental rotation around *X*-axes

When beginning in AP, segmental rotation around a joint's x-axis results in sagittal plane motion since the x-axis of the joint is parallel to the body's mediolateral axis. Consider the knee joint, for example. When the lower leg rotates around its x-axis starting from AP, the direction with the largest range of motion is for the knee to bend such that the foot moves posteriorly. On the other hand, rotating the forearm around the elbow joint's x-axis results in the largest range of motion when the hands move anteriorly. In general, *flexion* is the term given to joint action around an x-axis moving in the direction from AP toward the largest range of motion and *extension* is rotation around the x-axis in the direction back toward AP. The ankle joint is a special case that deviates from this standard terminology. *Dorsiflexion* is when the foot makes an acute angle with the lower leg (commonly called "flexed" in dance), and *plantarflexion* is when the foot makes an obtuse angle with the lower leg (a "pointed" foot in dance).

Hyperextension describes motion of the segments from AP toward the direction with lesser range of motion. Hyperextension at the knees, elbows, or other joints is not unusual in some dance forms. Studies have shown that dancers in certain styles, such as ballet, have greater hypermobility measures compared to non-dancers [95]. The opinion of some in the dance community is that hyperextension enhances aesthetics. Joint flexibility is fundamental in some forms of dance, and the lines created with hyperextended joints may be viewed as beautiful. While joint flexibility can be an important characteristic in some dance styles, taking it to the extreme of hypermobile joints may have negative consequences. Some studies have found an association between hypermobile joints and an increased risk for injury in dance [13]. More research is needed in this area, however, to understand whether hypermobility is a liability for dancers [24, 19].

2.2.2 Segmental rotation around *Y*-axes

Segmental motion in the frontal plane results from rotation around a joint's *y*-axis. For the shoulder and hip joints, *abduction* is rotation of the limb around its *y*-axis, starting from AP and moving in a direction away from the body's midline. *Adduction* is motion of the limb in the frontal plane back toward AP. For the head and trunk, frontal plane joint actions can be described by the terms *lateral flexion* to the left or right.

The ankle joint is again given its own terminology. In AP, the ankle joint axis parallel to the body's *Y*-axis is the foot's longitudinal axis. When rotation of the foot around its longitudinal axis makes the sole of the foot face medially, this is *inversion*. When the sole of the foot points laterally, this is *eversion*. A common technical flaw when "pointing" the foot in dance is to accompany the plantarflexion with inversion, often referred to as a "sickling" ankle.

2.2.3 Segmental rotation around *Z*-axes

Internal rotation and *external rotation* are terms to describe rotation of segments around a joint's *z*-axis, which in AP is parallel to the vertical. For example, external rotation at the hip occurs when the thigh segment rotates around its longitudinal axis and the knees point away from one another. Conversely, internal rotation at the hip causes the knees to point inward, toward each other. Proper ballet turnout is derived solely from hip external rotation. Due to the joint structures, the internal/external rotation range of motion is more limited at the knees and ankles than for a "ball and socket" joint like the hip. Still, some internal/external rotation is possible at the knees and ankles when there is enough friction between the feet and the floor. Forcing turnout from the knees and ankles, however, may place stress on the connective tissues of the joints and increase risk of injury. However, more research is needed to investigate turnout and injuries [41].

2.3 Human musculoskeletal system

The body's many systems all serve critical roles for human body functioning. These include, for example, the respiratory system, which brings oxygen from the outside environment into the body, and the cardiovascular system, which transports oxygen and other nutrients via blood to the many organs and tissues that need them. While these processes are vital to the functioning of our muscles and bones, the systems most directly involved in producing and controlling human movement are the musculoskeletal and nervous systems. Therefore, these are the biological systems discussed in more detail in this chapter.

As the name suggests, the musculoskeletal system includes our muscles and bones. It also consists of joints (which are much more than an abstract point of connection between bones) and connective tissues. Our bones provide structural support, protection for organs, and serve as levers powered by muscles to produce movement. In addition to generating motion, our muscles provide joint stability and allow us to breathe and our hearts to beat. Connective tissues include tendons (connect muscles to bones), ligaments (connect bones to bones), fascia (surrounds internal structures such as organs, muscles, bones, blood vessels, and nerves), and cartilage (e.g., in our noses and ears, a smooth and flexible tissue that serves multiple functions in the body). The bones, muscles, joints, and connective tissues together form the complex mechanical structure of the human body.

2.3.1 Bone and the skeletal system

Bone provides multiple important functions for the human body that are both physical and physiological. With their relatively rigid structure, bones provide physical framework for the body, can absorb energy from external loading, and offer protection for underlying soft tissues and organs. Bones also serve as stores of fat and minerals such as calcium and phosphorus that can be released to the body if needed. Bone marrow contains different types of stem cells that can produce other cells such as blood cells, bone cells, and muscle cells. In the biomechanical analyzes in this book, the mechanical functions of bone will be our primary focus.

Bones of the head, neck, and trunk region are considered parts of the *axial* skeleton and those in the upper and lower limbs constitute the *appendicular* skeleton (Fig. 2.2). The shapes of bones reflect important aspects of their function. For example, the flat bones of the ribs and skull in the axial skeleton provide protection for internal organs and the brain. The long bones of the appendicular skeleton, such as the femur in the thigh, are effective at withstanding large compressive loads, such as during landings from jumps or the percussive taconeo, or footwork, in Flamenco. Another important function of bones is that they act as levers in which the action of forces on the bone (e.g., muscle forces, gravitational forces) can produce motion about a pivot point. The coordinated movement of all the various levers leads to complex movement like dance! In this section, we will study the composition of bone and its material and mechanical properties to understand how bone structure relates to function and injury.

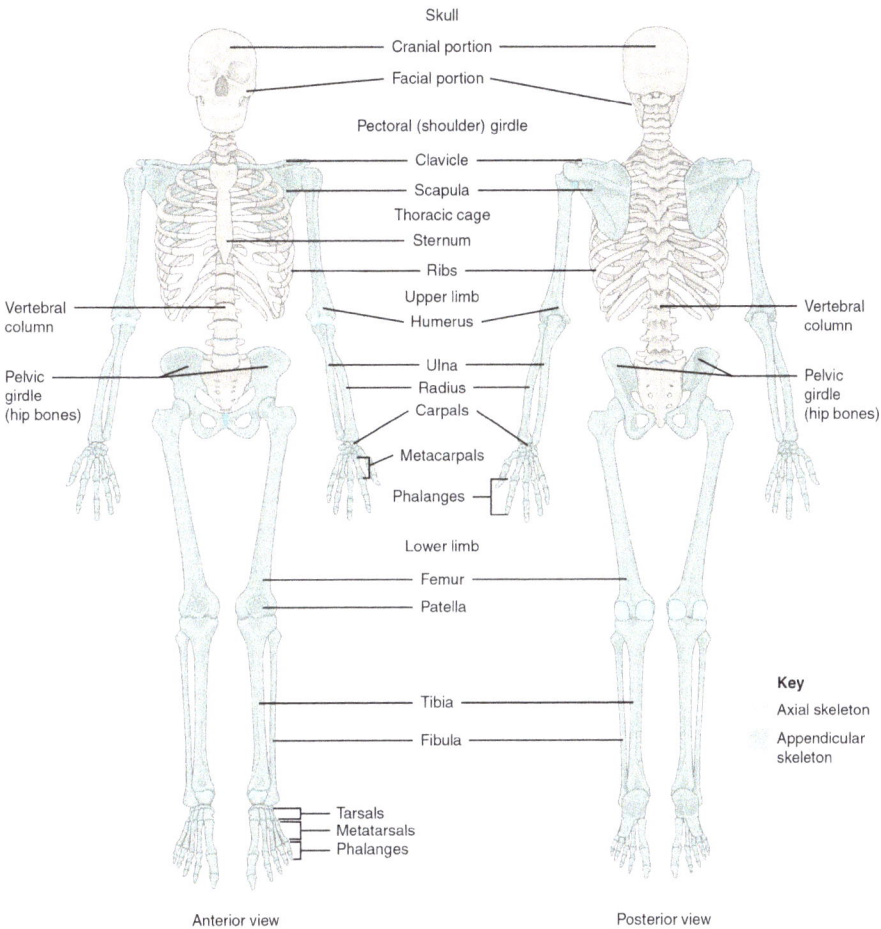

Figure 2.2: Anterior view (left) and posterior view (right) of the endoskeleton of a human adult. Axial and appendicular skeletal features are indicated. "Appendicular Skeleton" by "OpenStax" is licensed under CC BY. Access for free at https://openstax.org/books/anatomy-and-physiology/pages/1-introduction.

Bone is composed of inorganic material (~40 % by volume), organic material (~35 %), and water (~25 %). The inorganic material is an impure type of calcium phosphate and the organic material mainly consists of forms of collagen. *Collagen* is the strongest and most abundant protein in the body, found not only in bones, but other structures including tendons and ligaments. These organic and inorganic materials are arranged into a mineralized collagen fibril ~1 micrometer long that is considered the basic building block of bone tissue. These fibrils are organized in different ways to form the larger bone tissue structures and ultimately the entire bone. The two primary types of bone tissue are *cortical (compact) bone* and *cancellous (spongy) bone*. Cortical bone is denser with more structured arrangement of fibrils than cancellous bone. It is located in the shaft of long bones and forms the hard outer layer of bones (Fig. 2.3), providing stiffness and strength to the bone structure. Cancellous bone is less dense, porous bone and is found beneath the cortical bone at the ends of long bones, helping to absorb loads.

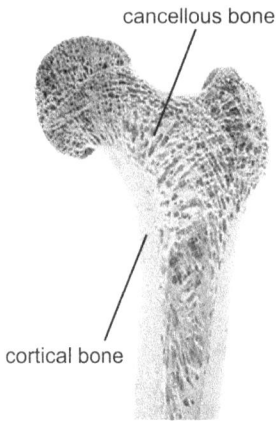

cancellous bone

cortical bone

Figure 2.3: Coritical and cancellous bone in a human femur. Reprinted with permission from "The microstructural and biomechanical development of the condylar bone: A review", European Journal of Orthodontics 36(4), 2013. DOI:10.1093/ejo/cjt093.

During physical activities, our bones can withstand large and repetitive forces. Most often in dance a single force on a bone does not exceed the bone's overall breaking strength; however, repetitive loading can still damage the bone and lead to overuse injuries like stress fractures. It has been shown that ballet dancers who dance for more than five hours per day are more likely to develop stress fractures than those who dance for less time, and the most common location for stress fractures for these dancers was the metatarsals (in the feet) followed by the tibia (shin) [40].

Because bones provide structural support for the body, it is important to understand their mechanical properties and how bones behave when loaded. Mechanical properties such as strength and elasticity depend on multiple factors, including the material properties of the underlying bone tissue (cortical or cancellous bone), how the material is organized to form the overall structure, and the geometry of the whole bone.

Material properties of the bone tissue itself can be measured experimentally. For example, a small piece is taken from the whole bone, and measurements of how much the bone deforms (e. g., compresses or stretches) under known loads are collected. *Stress* is defined as the force per unit area (in units of N/m^2 or Pascals (Pa)) the tissue experiences,

$$\text{Stress} = \frac{F}{A}$$

and *strain* is a unitless quantity measuring the fractional change in length of the bone (relative to its original length L_0)

$$\text{Strain} = \frac{\Delta L}{L_0}$$

If bone experiences a tensile force, it stretches (or lengthens), even if ever so slightly. Under a compressive force, bone's length decreases. When the applied forces are small enough, the bone behaves in manner such that the force required to stretch or compress the bone increases proportionally with its change in length. Similar to Hooke's Law for an ideal spring familiar to students of physics ($|F| = k|\Delta x|$), stress and strain are linearly proportional in what is referred to as the elastic region for bone.

$$(\text{Stress}) = Y(\text{Strain}) \tag{2.1}$$

A plot of stress vs. strain in this region produces a straight line, and the constant slope Y is called *Young's modulus* (with the same units as stress (Pa), since strain is a unitless quantity). Young's modulus quantifies how easy or difficult it is to deform the material. A material with a larger Y is more resistant to deformation.

When stress is relatively small, the bone returns to its original length when the load is removed. However, when the stress exceeds the elastic limit, the bone becomes permanently deformed, no longer returning to its original length when the load is removed. If the stress continues to increase, eventually the bone reaches its breaking point (fractures).

Fig. 2.4 shows stress-strain curves for cortical bone tissue and cancellous bone tissue under deformation. As shown in the graph, the denser cortical tissue can withstand greater overall stress than the cancellous tissue, but has lower maximum strain (relative deformation) before fracture. The plot shows that cortical bone's Young's modulus (the slope of the line in the elastic region (before σ_{yield})) is greater than cancellous bone, meaning that the dense cortical bone is stiffer than the spongy cancellous bone. There is of course variation in bone tissue material properties depending on multiple factors (e. g., age, gender, mineral content), so measures such as Young's modulus are often reported with large ranges.

A large factor affecting the mechanical properties of bones is the porosity of the bone. Because different bones, or even different locations within the same bone, have different porosities, they have different overall mechanical properties. The orientation and organization of the fibrils leads to bone having different properties depending on the direction of loading (compression/tension, shear, or torsion). For example, the arrangement of cortical bone along the length of long bones (like wood grain) leads to them being strongest in the longitudinal direction, such that they can withstand large compressive forces, good for running and jumping. Most fractures are due to bending, torsion, or a combination. Another important factor to consider that affects mechanical properties is the rate of loading. If bone is loaded rapidly, it is stronger. For slower, viscoelastic loading bone is weaker.

Although we often think of our bones as being relatively unchanging structures, bones are very much alive and everchanging. Even the bones of a fully grown, adult body continuously change through the processes of *resorption* (bone tissue absorption) and *ossification* (bone tissue growth). Specialized cells, called osteoblasts, produce new bone during ossification by excreting collagen and controlling the inorganic material

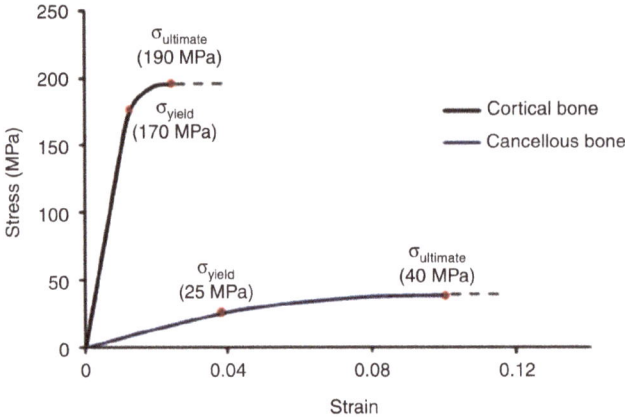

Figure 2.4: Stress vs. Strain for cortical bone tissue and cancellous bone tissue. (Note that σ_{yield} is the elastic limit, and $\sigma_{ultimate}$ is the bone's breaking point). Reprinted with permission from "A comparative study of tapped and untapped pilot holes for bicortical orthopedic screws – 3D finite element analysis with an experimental test", Biomedical Engineering/Biomedizinische Technik, vol. 64, no. 5, 2019, pp. 563–570. https://doi.org/10.1515/bmt-2018-0049.

deposited in bone. Osteoclasts, on the other hand, dissolve the bony matrix during resorption by excreting acids and can release the stored minerals to the body. A proper balance between resorption and ossification is needed for bone health.

Certain factors affect the rate at which bone tissue is created and resorbed. Bone changes its structure, size, or shape over time, based on factors such as age and mechanical demands. While extreme mechanical demands can result in negative consequences such as bone fracture, less extreme loading is not only tolerable, but it is necessary for the proper balance between resorption and ossification. If there are long periods of time when the bones experience very little stress (e. g., astronauts in reduced gravity environment), more bone tissue will be resorbed than created and the bones can become weaker. Loading experienced by bones while dancing has the potential to promote bone heath and improve bone mineral density. Studies have found bone density to be in the normal or elevated ranges in weight bearing locations (e. g. legs of ballet dancers) but lower than normal in non-weight bearing locations (e. g., arms of ballet dancers) [105, 111]. However, forms of dance like ballet and other sports that endorse very low body weight have been linked with low bone mineral density (BMD) due to low food energy intake and/or extreme energy expenditure [102]. Short- and long-term health consequences of low BMD in dancers are unclear.

Despite all this discussion of the properties of living, changing bone, the relative rigidity of bone allows for us to model body segments as rigid bodies in many applications. Tissue mechanics is an important subfield of biomechanics, but for the functional mechanics of human movement analysis, we will primarily model bones as rigid levers connected at joints. The next section covers this important intersection between bones and the structure and function of joints.

2.3.2 Joints

A *joint*, or articulation, is the term commonly utilized to describe where bones are connected. Some joints are rigid, like the fibrous joints of the skull bones; our primary focus will be on a much more mobile type of joint, synovial joints, which comprise most joints of the appendicular skeleton. Joints themselves are not an idealized point or simply the location where the bones join, but are physical structures between and connecting the bones. In a *synovial joint*, ligaments connect bone to bone and a joint capsule of fibrous tissue encloses the small space between the adjoining bones (Fig. 2.5). This space (joint cavity) is not empty, but is filled with low-friction synovial fluid, allowing the bones to easily move and rotate relative to one another. The articulating surfaces of the bones are covered by cartilage, which acts as a shock absorber for the bones. The articular cartilage secretes synovial fluid when loaded, so the joint is self-lubricating. In some joints extra separation and shock-absorption between bones is provided by separate disc(s) of cartilage (e. g., knee meniscus).

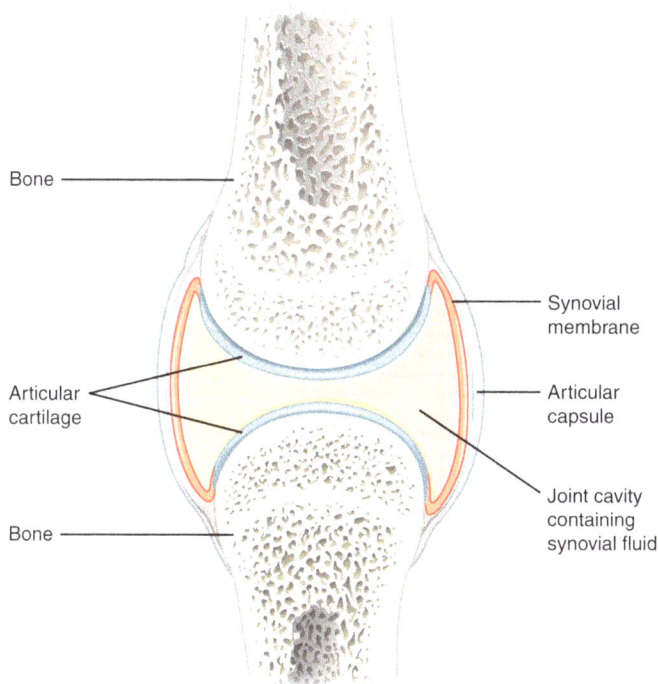

Figure 2.5: "Synovial Joint" by "OpenStax" is licensed under CC BY. Access for free at https://openstax.org/books/anatomy-and-physiology/pages/1-introduction.

Different joints allow or restrict movement of the adjoining segments depending on their structure. The shape of the bones and how they fit together at a joint influence what motion is or is not allowed. There are six main types of synovial joints (Fig. 2.6): ball and socket, hinge, pivot, gliding (plane), saddle, and ellipsoidal (condyloid).

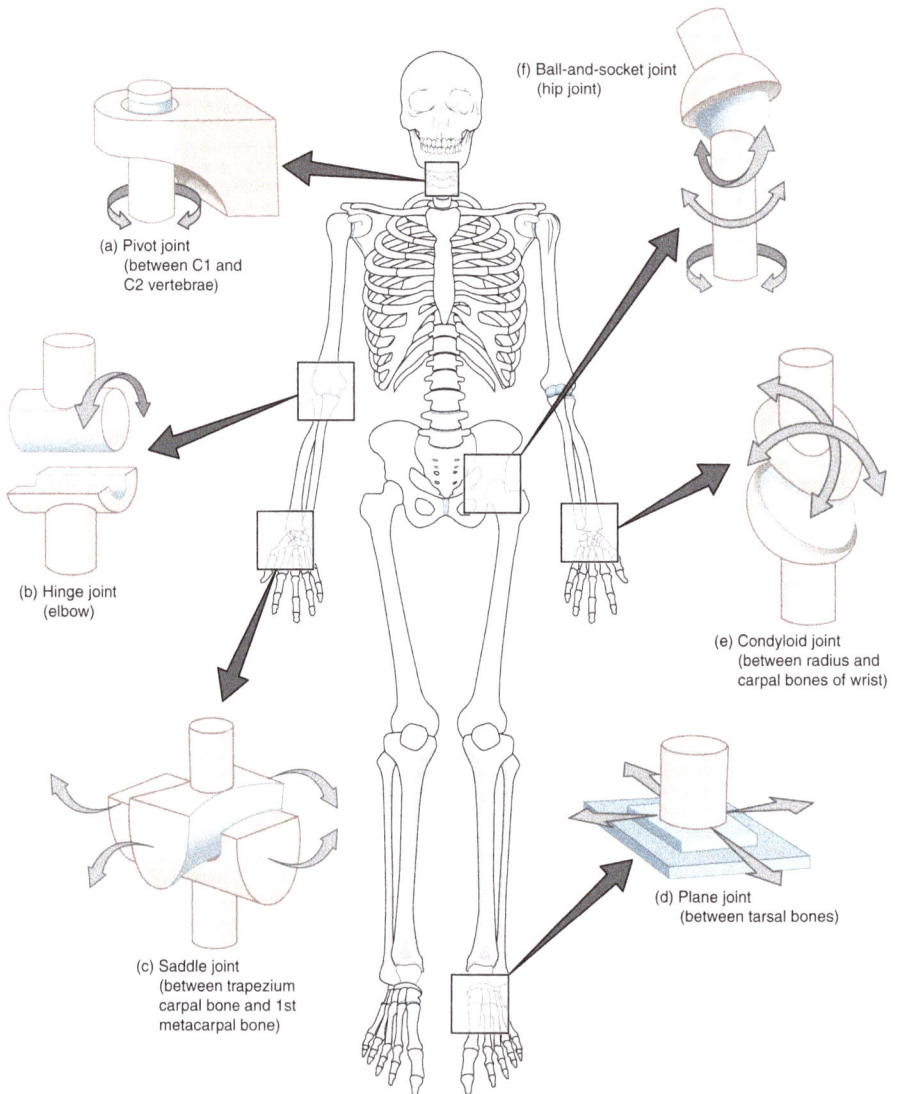

Figure 2.6: Types of synovial joints: (a) pivot, (b) hinge, (c) saddle, (d) plane (gliding), (e) condyloid (ellipsoidal), and (f) ball and socket. "Types of Synovial Joints" by "OpenStax" is licensed under CC BY. Access for free at https://openstax.org/books/anatomy-and-physiology/pages/1-introduction.

In a *ball and socket joint*, the rounded end of one bone ("ball") fits into a spherical cavity of the adjoining bone ("socket"). For example, the spherically shaped head of the femur fits into the cavity of the pelvis (acetabulum) in the hip joint. Ball and socket joints allow for the maximum range of motion in all three rotational degrees of freedom. Another ball and socket joint that allows for large rotational movement is the glenohumeral joint of the shoulder. In this case, the head of the humerus is the ball that fits into the socket of the scapula (glenoid cavity).

If the shape of the "ball" and "socket" are changed such that they are egg shaped (ellipsoidal), the joint type is instead an *ellipsoidal joint* (also known as a condyloid joint). The egg shape impedes axial rotation, reducing the rotational degrees of freedom from three to two. The metacarpophalangeal joints (where the fingers meet the hand) are ellipsoidal. The fingers can flex and extend or ab/adduct, but they cannot rotate around the longitudinal axis.

The articulating bones of a *saddle joint* can be compared to a rider sitting on a saddle. One bone has a concave shape (the saddle), and the other bone's convex shape (rider) fits into it. The motion of a saddle joint is similar to the ellipsoidal joint in that it allows for two rotational degrees of freedom, although generally the range of motion at a saddle joint (e. g., thumb) is greater.

A *hinge joint*, like a door hinge, primarily allows for just one rotational degree of freedom (flexion/extension). Two examples of hinge joints are the elbow and knee joints. Another type of joint allowing only one rotational degree of freedom is a *pivot joint*, but in this case the rotation is axial. One bone fits into a circular opening on the adjoining bone and the rotation occurs along the first bone's longitudinal axis. A pivot joint between the first and second cervical vertebrae allows us to turn our heads to look left and right.

Finally, in a *gliding joint* (also known as a plane joint), the articulating surfaces of the bones are nearly flat or planar, which allows the surfaces to translate (or glide) relative to one another. This allows for translational degrees of freedom (as opposed to rotational degrees of freedom) between the bones, such as in the acromioclavicular joint of the shoulder. Typically, the translational movements are small, being limited by the ligaments crossing a joint.

While dancing, and during human movement in general, there is a balance between joint stability (preventing motion outside of the joint's normal planes of motion or dislocation of adjoining surfaces) and joint flexibility (large range of motion in the appropriate plane(s) of motion). *Range of motion* (ROM) at a joint is limited by the structures surrounding the joint (bones, muscles, tendons, and ligaments), which are also structures that can help provide stability to the joint. The extensibility of a muscle is affected by the muscle's length, how extensible its fibers are, as well as whether the muscle is biarticular (if it spans two different joints). The hamstrings are biarticular, crossing behind both the knee and the hip joints. When the knee is extended and the hip flexed, the hamstrings are lengthened by both of these actions. The result is a decrease in hip flexion ROM when the knee is extended, and hip flexion ROM increases when the knee is flexed.

Large ROM at certain joints is needed to meet the artistic demands of many dance styles. Dancers are challenged with the task of increasing ROM at these joints without sacrificing strength or stability (both joint stability and postural stability). Stretching exercises can increase ROM, but research has shown that not all types of stretches are beneficial for all situations. Research is ongoing to understand the effects of static stretching, active stretching, and other methods of stretching on dance performance. A summary and recommendations for dancers can be found in "Stretching for Dance" by Matthew Wyon in The IADMS Bulletin for Teachers (2010).

2.3.3 Muscles

Of the three different types of muscle in the body (skeletal, cardiac, and smooth), our focus will be on skeletal muscles, which can activate under voluntary control from the nervous system. The other two types of muscle are involuntary: the cardiac muscle of the heart and smooth muscle that, for example, lines our digestive tract. The primary function of skeletal muscles is to exert forces on the body's segments through their attachment to bones via tendons. In engineering terms, they can be thought of as the body's actuators. Skeletal muscle can only actively contract or shorten, and muscles can only pull (not push) on the segments to which they are attached (like tension force).

Muscle forces can provide a net torque on segments and induce acceleration of body parts. They can also prevent unwanted motion at a joint like ensuring that the knees do not collapse during standing. In this section, we will discuss the functional roles of muscles, muscle actions, the basic structure of muscle, and factors that affect the force muscles can produce under different static and dynamic conditions.

2.3.3.1 Functional roles of muscles

The muscles spanning a joint can take on a variety of roles in producing or preventing movement. The role taken on by a particular muscle or muscle group is also situation dependent. Even a seemingly isolated movement like a front kick (hip flexion) requires the activation and coordination of multiple muscles spanning a joint or joints. In general, agonists are the muscles that produce the joint action. *Agonists* exert a torque that is in the same direction as the angular velocity between limbs. *Antagonists* cross a joint on the opposite side as the agonists and exert torque in the opposite direction of angular velocity. For example, the biceps and triceps are an agonist/antagonist pair during elbow flexion (Fig. 2.7). With the agonists and antagonists simultaneously activated (termed *coactivation* or cocontraction), the movement can be more refined and controlled.

There are also muscles that are not primary actors in the movement, but that play important supporting roles. Muscles that help hold the joint together are *stabilizers*. These are often deeper muscles, close to the joint, like the rotator cuff of the shoulder that helps to hold the head of the humerus within the relatively shallow cavity within

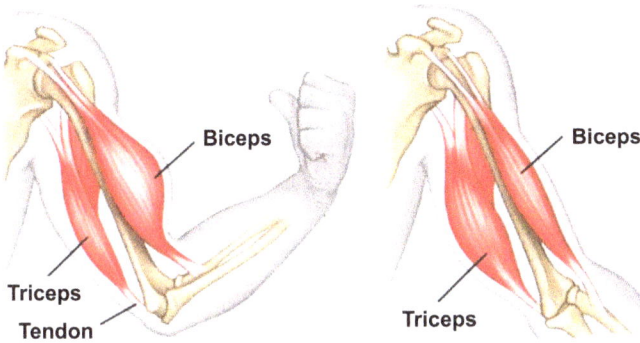

Figure 2.7: The biceps is the agonist and triceps the antagonist during elbow flexion (left). During elbow extension, the triceps is the agonist and the biceps is the antagonist (right). Reprinted with permission from "Static analysis of a torsion motor generating flexion – extension motions of the elbow" MATEC Web of Conferences 178, 07005 (2018) https://doi.org/10.1051/matecconf/201817807005. Licensed under CC BY.

the scapula. A muscle that acts as a *fixator* helps to prevent undesired motion of a body segment when other muscles are active. For example, grand battement devant, a type of front kick in ballet, requires hip flexion where the lifted leg is the only part of the body that noticeably moves. The agonists in this situation are a group of muscles, called the hip flexors (iliopsoas, rectus femoris, sartorius), that in general tend to rotate both segments to which they are attached: the thigh (femur) and the pelvis. In order to prevent unwanted downward tilting of the pelvis during grand battement devant, the abdominal muscles (rectus abdominus) act as fixators to the action of the hip flexors. The agonists also in general produce motion that may be outside of the desired plane of motion, so other muscles can act as *neutralizers* to counteract the unwanted motion. For example, during hip extension, the primary agonist is gluteus maximus. However, gluteus maximus when activated alone produces both hip extension and external rotation. To counteract the external rotation, the internal rotators of the hip can act as neutralizers. In general, many muscles act together to produce and control the desired movement.

2.3.3.2 Muscle actions

When a muscle is "activated," muscle tension is actively generated, and the force is transmitted to the bones to which it is connected via tendon. Muscle actions are classified as isometric, concentric, or eccentric depending on how the length of the muscle changes during its activation. During *isometric activation*, the muscle's length remains constant while it is activated. Imagine holding a heavy weight steady with 90 degrees of elbow flexion. The biceps isometrically activates to hold the object at rest. *Concentric activation* occurs when the activated muscle shortens in length, whereas *eccentric activation* occurs when the activated muscle increases in length. Consider again holding the heavy weight. During elbow flexion when the weight moves upward and the angle between the forearm and upper arm decreases, the biceps undergoes concentric activation. When the weight is moved back down with elbow extension increasing the angle between the forearm and upper arm, the biceps is now lengthening (but still active), so this is an eccentric activation. (If you place your opposite hand on the biceps near your elbow when doing a bicep curl, you can feel the biceps shortening during flexion and lengthening during extension.)

2.3.3.3 Structure of muscle

In our discussion of muscles up to this point, we began with the fact that muscles produce force, but we have not yet questioned *how* they produce force. To understand the mechanics of muscle force production requires knowledge of the underlying structure of muscles. Whole muscles are composed of muscle cells (called muscle fibers), connective tissues, nerves, and blood vessels. The connective tissue exists both outside of the muscle body and within the muscle, separating muscle fiber bundles. The connective tissue is stiffer than the muscle fibers and contributes to the muscle's passive tension (or force it would provide when turned off but stretched).

The muscle fibers are long, cylindrical cells with a diameter of ~10–100 micrometers. They are composed of smaller myofibrils that contain the most basic contractile unit of muscle or the *sarcomere*. In the sarcomere are two overlapping proteins, myosin and actin, that bind together when the muscle is activated. These bonds, or "cross-bridges," create tension pulling in a direction that would decrease the length of the sarcomere and cause the actin and myosin filaments to slide relative to one another (Fig. 2.8). This model of muscle activation is known as the sliding filament theory. Sliding filament theory can help us understand why muscle force changes depending on the length of that muscle.

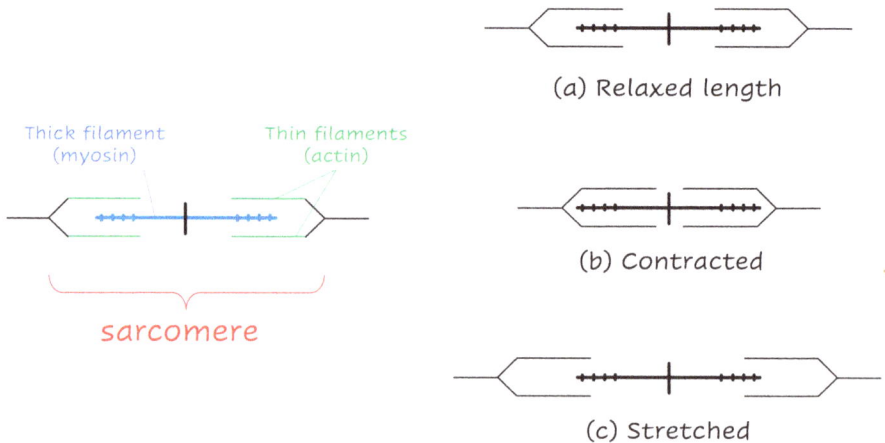

Figure 2.8: Schematic of the thick (myosin) and thin (actin) filaments of a sarcomere at a muscle's (a) relaxed, (b) contracted (shortened), and (c) stretched length.

2.3.3.4 Muscle length and force

The maximum force that a muscle can produce depends on the muscle's length relative to its relaxed length. We can understand this by considering the cross-bridges formed during *isometric* muscle activation. Increasing the number of tension producing cross-bridges increases the force actively generated by the muscle fibers. For a single muscle fiber at its relaxed length (neither stretched nor contracted), the greatest number of cross-bridges can be formed due to the overlap in the actin and myosin filaments. When the muscle fiber shortens, the number of possible sites for cross-bridge formation decreases. When the muscle fiber is lengthened past its resting length, less of the thin filament overlaps with the thick filaments, again decreasing the number of cross-bridges. Therefore, as length of the muscle increases or decreases past its resting length, the amount of active force it can produce also decreases.

While it is true that muscle fibers are the part of the whole muscle that actively generate force, even a muscle that is "turned off" can exert a force passively due to the tension in the muscle and connective tissues. The muscle-tendon unit has some elasticity that we can think of somewhat like a spring, increasing the more that the spring is stretched past its resting length. So, the relationship between total muscle force and length for a whole muscle includes both active and passive contributions (Fig. 2.9).

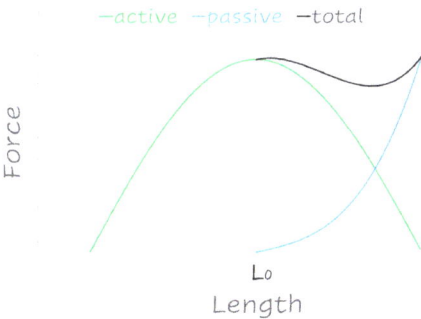

Figure 2.9: Force vs. length of a muscle, including connective tissue. L_0 = relaxed length. The blue line (passive) is the force vs. length of the resting (inactivated) muscle, primarily due to the elasticity of the connective tissue. The green line (active) is the force generated by the activation of the muscle fibers. The black line is the total muscle force when stretched past its resting length (sum of active and passive).

2.3.3.5 Muscle force–velocity relationship

In the previous section, we considered how the length of the muscle affected its ability to produce force during *isometric* activation. When there is movement, however, the length of the muscle changes *throughout* the activation. The velocity at which the muscle shortens or lengthens also affects the amount of force the muscle can produce (Fig. 2.10). We can understand this by again considering the ability of cross-bridges to form. When the thin and think filaments are sliding relative to one another, it is more difficult for the myosin protein to "catch" a binding site on the actin to form a cross-bridge. As the velocity of muscle shortening increases during concentric activation, the cross-bridge formation becomes more and more inefficient, producing less overall muscle force. On the other hand, if the muscle is lengthening (eccentric activation), we must again consider the elastic elements of the muscle. As the speed of the eccentric activation increases, muscle force increases.

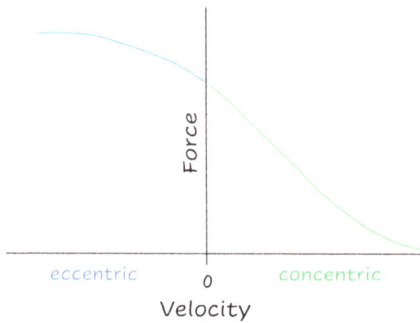

Figure 2.10: Force-velocity relationship for muscle. Positive velocity indicates the muscle is shortening (concentric activation), and negative velocity is lengthening (eccentric activation). The more rapid the shortening, the less force the muscle can produce, whereas increasing the speed at which the muscle lengthens increases muscle force.

2.4 The nervous system and motor control

Muscle activation, whether voluntary or involuntary, is initiated and controlled by the nervous system. The central nervous system (CNS) consists of the brain and spinal cord and is where sensory feedback is integrated and top-down commands originate to control motor output. The peripheral nervous system (PNS) is an information transfer system consisting of neurons leading between the CNS and the rest of the body. Some neurons are dedicated to regulating our body's automatic functions (e. g., digestion, cardiac functioning), while others innervate skeletal muscle. *Motor neurons* carry electrical signals from the CNS to activate muscles. A *motor unit* consists of one motor neuron and the muscle fibers to which it is connected (Fig. 2.11). The fibers of a motor unit are not localized, but spread throughout the muscle. The number of muscle fibers in a motor unit varies from as small as around ten to greater than one thousand. More refined muscle control is possible with smaller number of muscle fibers per motor unit.

When an electrical impulse called an *action potential* is sent from the motor neuron to the muscle fibers, they "contract" or activate. The action potential is an all-or-nothing signal telling the muscle fibers to turn on. The muscle response to a single action potential is referred to as a twitch. There is some lag time (latency) between when the

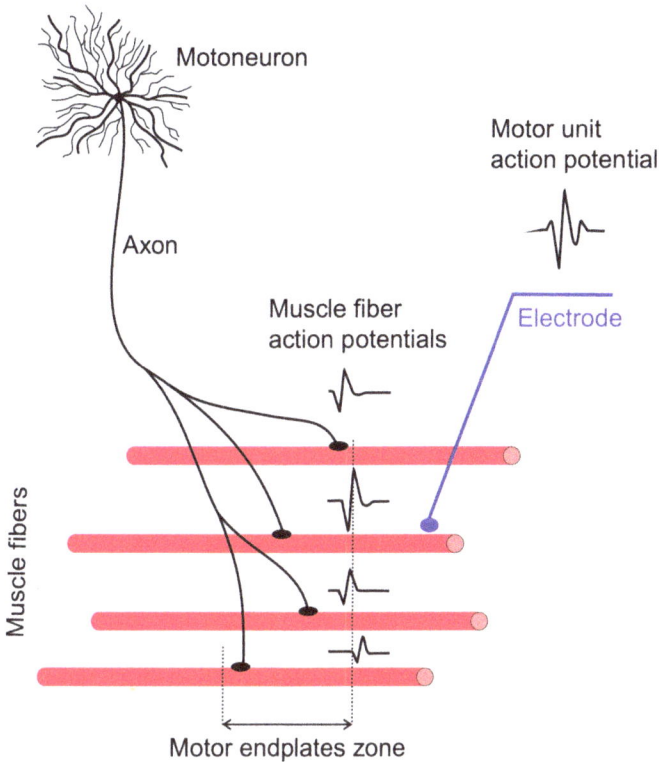

Figure 2.11: A single motor unit. Reprinted with permission from Jan Celichowski, Piotr Krutki, Chapter 4 – Motor Units and Muscle Receptors, Editor(s): Jerzy A. Zoladz, Muscle and Exercise Physiology, Academic Press, 2019, pp. 51–91, ISBN 9780128145937, https://doi.org/10.1016/B978-0-12-814593-7.00004-9.

electrical signal is received and when the muscle begins to generate tension that varies between different muscles, but is in the range of a few milliseconds. The total duration of the twitch also varies between muscles. A single twitch generates a fixed amount of tension in the muscle, but if another action potential reaches the muscle fibers quickly enough, the two (or more) twitches can add together, producing greater overall muscle tension (Fig. 2.12).

In addition to increasing the frequency at which action potentials are sent to a single motor unit, another way to increase the total amount of force generated by a single muscle is to increase the number of motor units recruited. This increases the total number of muscle fibers within that muscle that are actively generating force.

2.4.1 Energy for muscle activation

Even if action potentials are sent continuously to muscle fibers, there is a limit to the force that can be produced. This is because it takes chemical energy to create the bonds

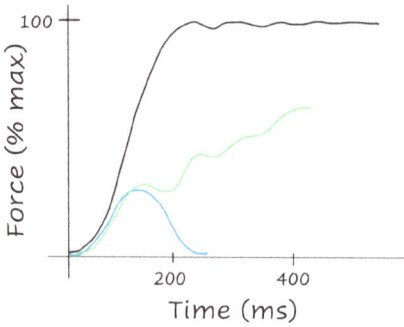

Figure 2.12: Increasing the frequency at which action potentials are sent to the muscle increases the force the muscle can produce, until it reaches its maximum, called "tetanus". (Blue = single twitch, Green = repeated stimulation, Black = tetanus.)

forming the cross-bridges in the sarcomere. If the muscle is stimulated for too long, the necessary energy sources can become depleted. There are two primary chemical processes to provide energy for muscle activation: (1) anaerobic metabolism of glycogen (sugar) and (2) aerobic metabolism, which requires oxygen. During less intense activity, enough oxygen is present for aerobic processes, but as the intensity of the activity increases, the oxygen stores may dwindle, so our muscles need to go to a different source of energy (glycogen). During anaerobic processes, lactate builds up in the muscles as a byproduct.

Muscle fibers are classified into three different types that differ in terms of their primary energy source as well as other characteristics such as the rate at which they respond to action potentials. Commonly referred to as "fast twitch" or "slow twitch" muscles, there are in fact three types (I, IIA, and IIB) with some of their main properties summarized in Tab. 2.1.

Table 2.1: The three skeletal muscle fiber types (Type I, Type IIA, and Type IIB) and their relative characteristics.

	I (Slow Oxidative)	IIA (Fast Oxidative)	IIB (Fast Glycolytic)
Activation time	Slow	Fast	Very Fast
Resistance to fatigue	High	Medium	Low
Force	Small	Medium	Large
Size of Fibers/Motor Neurons	Small	Medium	Large

How much of each fiber type an individual has seems to depend primarily on genetics, however, there is some evidence to suggest that muscle fiber composition can be changed with training. Type I fibers are best for activities that have a long duration and require smaller forces (e. g., long-distance running). Type IIB fibers can exert large forces in short bursts but are susceptible to fatigue because they rely on anaerobic metabolism. These are the fast-twitch muscle fibers that are useful during high intensity activities like sprinting. Type IIA are something in between the other two types and they can use both aerobic and anaerobic processes. Type IIA fibers can therefore be used in either high-intensity or sustained activities.

2.4.2 Summary of factors contributing to total muscle force generation

We have seen that the total active tension that a muscle can produce fundamentally depends upon the number and strength of the cross-bridges formed within each sarcomere. Factors that are physical, biochemical, and neurological impact cross-bridge formation. In addition to the active force generated via the cross-bridges, passive tension from the stiffness of the connective tissues contributes to total muscle force. To summarize, the total force a muscle exerts depends upon:
-   Composition of the muscle (muscle fiber type(s) and stiffness of connective tissue).
-   Muscle length.
-   Velocity of the muscle's change in length.
-   Number of motor units recruited.
-   Frequency at which motor units are stimulated by action potentials.

2.4.3 Stretching and neuromuscular function

Stretching is a regular part of many dancers' practice, often as part of a warmup, and is utilized often with a goal of increasing the range of motion at dancer's joints. Joint flexibility is an essential criteria for some forms of dance. There are some potential drawbacks to certain types of stretching however that dancers should be aware. Some studies have found decreases in muscular performance after stretching, but not all stretching is created equal. Stretches held statically for long periods of time can decrease muscle-tendon stiffness and affect the force–velocity and force–length curves for a muscle, overall decreasing its force production capabilities. Static stretching may also decrease the number and frequency of action potentials sent to the muscle. *Dynamic stretching* of muscles, in which a dancer actively moves a joint through its full range of motion, though less effective at promoting range of motion at a joint, can lead to increased muscle activation and is beneficial for movements that require large forces or power [72]. Overall, for the varied demands of dance performance, a combination of moderate duration static stretches and dynamic stretches is likely most effective.

2.4.4 Sensory information and control strategies

Effective and safe dancing is not achievable without the help of sensory information. The human body is a complex system with many segments and muscles to coordinate.

The environment can change suddenly while dancing – uneven flooring or changes in friction, unexpected proximity to a partner, etc. – so it is important to continuously monitor and refine output commands based on sensory stimuli.

Sensory neurons carry sensory information from the body back to the CNS. Of the five basic senses you may have learned about as a child – smell, taste, sight, hearing, and touch – the latter three are the most relevant for dance. A dancer sees where they are in space and in relation to other dancers and adjusts where they move, they hear the music and speed up or slow down to keep tempo, and they feel a partner's hand or how their feet contact the floor. However, sorting the senses into these five categories is overly simplistic, particularly for touch. Not only can we sense the feeling of contact between our skin and other surfaces, we feel our body in motion. *Proprioception* is our sense of the position and movement of parts of the body. Studies consistently show that as dance training increases, dancers rely more and more on proprioception as a feedback mechanism.

The sensory neurons that send information to the CNS receive this information from specialized sensory receptors. A subset of these sensory receptors important for motor functioning are the proprioceptors. *Proprioceptors* are located within various parts of the musculoskeletal system, such as muscles, tendons, and joint capsules, and provide us with perception about movement. Proprioceptors can also initiate involuntary reflexes with little to no input from higher order CNS, such as the simple "knee-jerk" reflex that is initiated by tapping the tendon just below the knee. Two important types of proprioceptors (muscle spindles and golgi tendon organs) are described below.

Muscle spindles are embedded within muscle, and detect changes in length of the muscle. As the muscle fibers are stretched, the muscle spindle stimulates a sensory neuron. This sensory stimulus depends on the velocity of stretch, with a rapid stretch leading to an increased stimulus. A rapid stretch can also lead to what is called the stretch reflex. In the stretch reflex, the stimulus from the muscle spindle travels along a sensory neuron to the spinal cord and then along a motor neuron back to the stretched muscle, causing it to activate and prevent further stretching. The sensory stimulus also acts to inhibit the motor neurons of the antagonistic muscle, an effect called "reciprocal inhibition." Activation of the stretched muscle and reciprocal inhibition of its antagonist combine to prevent excessive muscle stretch, which may otherwise lead to tearing of the muscle fibers.

Another proprioceptor, the *golgi tendon organ*, is located in tendon, close to the muscle, and senses tension force. As the force in the tendon and its attached muscle increase, the stimulus sent from the golgi tendon organ's sensory neurons increase. The result of the sensory feedback from the golgi tendon organ is to send an inhibitory response to the motor neurons of the muscle. If the muscle-tendon complex experiences excessive tension force, golgi tendon organ initiates a tendon reflex, causing the muscle to relax. The tendon reflex acts to prevent the muscle and/or tendon from tearing or from pulling away from their attachments to bone.

Whether and how the CNS integrates information sent from sensory receptors to impact motor output depends on multiple factors. Theories of motor control and motor learning are complex, ever evolving, and beyond the scope of this book. There are, however, three simplified classes of motor control systems worth discussing: open-loop, closed-loop feedback, and feedforward control. In *open-loop control*, sensory information is ignored, and a top-down command is sent from the CNS generating the muscle activation patterns. This type of control often occurs in rapid (or "ballistic") movements with large accelerations. In this case, there is not enough time for sensory signals to modify the motor output. In *closed-loop feedback control*, the output (dance movement) is monitored by the sensory system, and the information is sent back to the CNS. This sensory feedback is then used to modify the motor commands sent from the CNS to the muscles. If there was a disturbance during the movement's execution, it can be modified after comparing the actual movement to the desired movement. If a dancer feels their torso leaning too far to the right when they are meant to hold it vertically during a slow movement, a motor command can be sent to the abdominal obliques to make the correction. *Feedforward* control uses sensory information, but prior to the execution of the movement. Sensory input is used for planning of the movement; for example, a dancer notices they are leaning further forward than usual in a slow movement prior to initiating a backward leap, so they adjust their muscle activation pattern to exert more force propelling them backward. Most dance movements are likely controlled by both feedforward and feedback systems.

3 Center of mass kinematics and dynamics

A frustrated dancer cannot understand why her feet keep landing on the ground noticeably prior to everyone else during a traveling leap choreographed to a slow piece of music. "How do I stay in the air for a longer time?" she wonders. The dancer thinks about trying to jump farther, attempting that method a few times, but without success. She then contemplates bending her knees more during the jump's preparation, or timing her arm movement differently, but she's also concerned about staying true to the demands of the choreography. "My dance teacher says I can't look like I am straining, and I still have to use correct technique," she thinks. Understanding a few basic principles of physics, in combination with muscle mechanics, could help this dancer learn how to achieve her goal without so many futile trial and error attempts.

In Chapter 2, we studied two *biological systems*, the musculoskeletal and nervous systems, and how they work together to produce human movement. We now extend our focus to study the body as a *physical system* subject to the laws of mechanics. In this chapter, the physical system of interest will be a dancer's entire body. A crucial first step in any physical analysis is to create an idealized representation, or physical model, of the system. Given the extensive time to scratch the surface of the structure and function of the body in the previous chapter, one could become discouraged when contemplating an effective way to create a physical model of something as complex as the human body. So why not start with the simplest physical model we can think of – a single point particle? You may wonder if modeling a dancer as a point particle would be anywhere near realistic or provide any valuable insight at all. At its core, dancing arises from the relative motion of all the parts of the body, so it is true that we will have far from a complete picture if we only consider translational motion of a point particle located at the dancer's center of mass (CM). However, as we will see, much valuable information can be uncovered from studying CM motion. Information such as what is or is not possible for the body to do based on the laws of physics, how the CM moves while different external forces act, or how a movement can be modified to decrease the magnitude of external force the dancer experiences and therefore reduce injury risk.

This chapter begins by defining the center of mass and how to locate and track the CM position of a moving dancer. Describing how the CM moves (kinematics), and why the CM moves the way that it does (dynamics) when external forces act on a dancer is central to this chapter. By the end of chapter, we will interpret and explain experimental results of a dancer's jumping data based on physical and biomechanical principles.

3.1 Center of mass

The *center of mass* (CM) of a physical system is a weighted average position of where the system's mass is located. For a system of two point particles with masses m_1 and m_2 at positions $\vec{x}_1$ and $\vec{x}_2$ (Fig. 3.1), the center of mass position is $\vec{x}_{CM} = \frac{m_1\vec{x}_1+m_2\vec{x}_2}{m_1+m_2}$. If $m_1 = m_2$,

https://doi.org/10.1515/9783110642292-003

the CM position is located at the midpoint between m_1 and m_2. If $m_1 > m_2$, the CM is closer to m_1 and vice versa. For N number of discrete point particles the CM position is

$$\vec{x}_{CM} = \frac{1}{M} \sum_{i=1}^{N} m_i \vec{x}_i \tag{3.1}$$

where $M = \sum_{i=1}^{N} m_i$ is the total mass of the system.

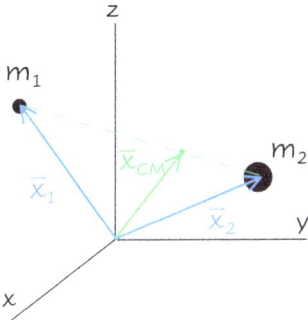

Figure 3.1: Center of mass position $\vec{x}_{CM}$ of two discrete point particles.

Example. With a mass of 1.59×10^{21} kg, Charon is Pluto's largest moon. Find the center of mass position of the two-body system of Charon and Pluto as a percentage of their total center-to-center distance, measured from the center of Pluto (mass of Pluto = 1.31×10^{22} kg).

Solution. Due to their spherical symmetry, we treat each celestial object as a point particle with its mass concentrated at its geometric center. For this example, we place Pluto and Charon on the x-axis with Pluto at the origin ($x_P = 0$, Fig. 3.2). Eq. (3.1) for this two-body system in 1-D becomes

$$x_{CM} = \frac{m_C x_C}{m_C + m_P} = \left(\frac{1.59 \times 10^{21} \text{ kg}}{1.59 \times 10^{21} \text{ kg} + 13.1 \times 10^{21} \text{ kg}} \right) x_C = 0.108 x_C$$

so the CM of the Pluto–Charon system is 10.8 % of the length from Pluto to Charon, closer to the more massive Pluto.

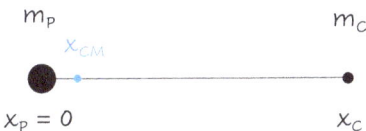

Figure 3.2: Two-body system of Pluto (m_P) and its moon Charon (m_C).

The human body consists of many more than two particles with mass distributed continuously throughout. In this case, we can think of the body as consisting of many infinitesimal bits of mass dm. To add up their contributions to the CM, the discrete sum in eq. (3.1) becomes an integral, $\vec{x}_{CM} = \frac{1}{M} \int \vec{x} \, dm$. The mass density for a three-dimensional object, $\rho = \frac{dm}{dV}$, can be used to perform the integration, so the CM position of a continuous mass distribution is

$$\vec{x}_{CM} = \frac{1}{M} \int \vec{x} \rho \, dV \tag{3.2}$$

where dV is the volume of the mass element (Fig. 3.3) (e. g., in Cartesian coordinates dV = dx dy dz).

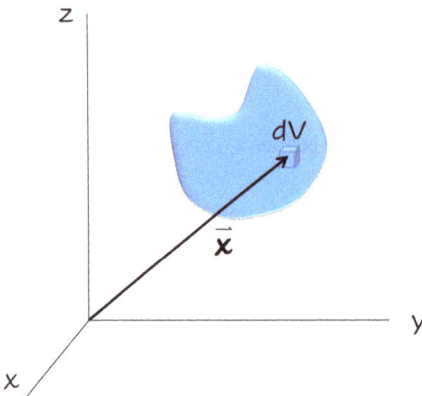

Figure 3.3: Continuous mass distribution with infinitesimal volume element dV at position $\vec{x}$.

If we imagine applying eq. (3.2) to find the center of mass of the entire dancer, we quickly encounter several practical problems. The human body is not a simple, well-defined geometric shape like a sphere, cone, or cylinder, so defining the boundary of the volume and performing the integration without many simplifying assumptions is foreboding. The density of the body also varies as a function of position, $\rho(\vec{x})$, as the infinitesimal bits of mass consist of parts of bone, muscle, fat, blood, etc. $\rho(\vec{x})$ is not in practice easily measurable or represented analytically by some function. Even if we experimentally measure the body's geometry and density with sophisticated body scans, the body's overall shape also changes substantially and rapidly while dancing as the body's parts move relative to one another. For example, if a dancer begins in anatomical position, then lifts the right leg to the side (hip abduction), the whole body CM also moves superiorly and laterally to the right.

Instead of using a single application of eq. (3.2) to compute the whole body CM, the previously mentioned challenges can be somewhat alleviated by implementing a *seg-mental analysis*, that is by separating the body into smaller, rigid segments that do not vary substantially in size or shape during movement. If the masses m_i and center of mass positions $\vec{x}_i$ of each individual segment are known, then the whole body CM can be computed from the discrete sum in eq. (3.1). Typical body segments for a whole body analysis are the head, trunk (torso), upper arms, forearms, hands, thighs, shanks (lower legs), and feet (Fig. 3.5).

Some of these body segments are better represented as a single segment (e. g., thigh and upper arm) because they are supported by one bone with joints at the segment endpoints. Other segments (e. g., foot and trunk) are much more complex and actually consist of multiple bones and joints throughout allowing the segments to change shape quite substantially depending on the movement. Many forms of dance have intricate foot or torso movements, so in some cases it may be more appropriate to use a multi-segment model for these body parts. Relatively little mass is contained in the feet, so the single-segment vs. multi-segment foot model does not have a substantial effect on the total body CM. How the massive trunk is represented can more substantially affect total body CM results.

Before applying the segmental analysis, we still encounter the challenge of knowing each individual segment's CM position and mass. Once we have chosen a rigid-body seg-ment with a certain size and shape, we may apply eq. (3.2) to determine the individual segment's CM position. For example, we could choose to model the thigh segment as a frustum of a cone with a larger proximal than distal diameter. These diameters and the length of the thigh segment can be measured directly for an individual dancer.

Example. Model a dancer's thigh segment as a frustum of a cone, where the radius of the distal end (knee) is R_D, the radius of the proximal end (hip) is R_P, and the longitudinal length is L. Assume uniform mass density ρ. Find the CM position of the thigh (measured from its proximal endpoint) in terms of R_D, R_P, L and constants.

Solution. Place the proximal end of the thigh at the origin with its longitudinal axis (the line connecting the hip joint center and knee joint center) aligned with the z-axis (Fig. 3.4). By symmetry, x_{CM} and y_{CM} are zero, i. e., the CM of the frustum lies along the longitudinal axis. We apply eq. (3.2) to find z_{CM}, where $\rho = \frac{M}{V}$, M is the thigh's mass, and V is its total volume:

$$z_{CM} = \frac{1}{M} \int z\rho \, dV = \frac{1}{V} \int z \, dV$$

We choose the infinitesimal volume element dV to be a thin disk with circular cross-sectional area $A = \pi r^2$ and infinitesimal thickness dz, so $dV = \pi r^2 \, dz$. The radius of the disk, r, varies with z, so the constant slope of the cone's edge (Fig. 3.4) can be used to find an expression for $r(z)$.

$$\text{slope} = \frac{\text{rise}}{\text{run}} = \frac{z}{R_P - r} = \frac{L}{R_P - R_D}$$

Solving for r we have $r = R_P + (R_D - R_P)\frac{z}{L}$ and the integral becomes

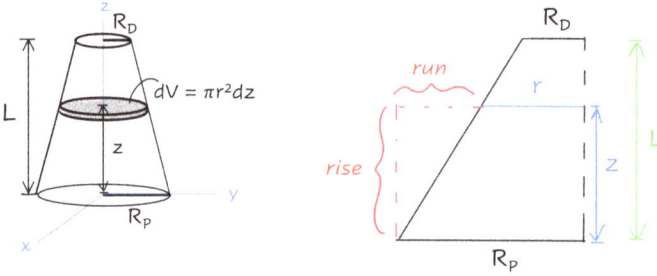

Figure 3.4: Frustum of a cone to represent a thigh segment. The longitudinal length is L, radius of proximal end (hip) R_P, and radius of distal end (knee) R_D.

$$z_{CM} = \frac{1}{V} \int_0^L 2\pi \left(R_P + (R_D - R_P)\frac{z}{L} \right)^2 dz$$

$$= \frac{\pi L^2}{12V} (3R_D^2 + 2R_D R_P + R_P^2)$$

The total volume of the frustum is found by integrating $V = \int dV = \int \pi r^2 \, dz$.

$$V = \int_0^L \pi \left(R_P + (R_D - R_P)\frac{z}{L} \right)^2 dz$$

$$= \frac{\pi L}{3} (R_D^2 + R_D R_P + R_P^2)$$

Substituting this expression for V into that for z_{CM}, we have our final solution.

$$z_{CM} = \frac{(3R_D^2 + 2R_D R_P + R_P^2)}{4(R_D^2 + R_D R_P + R_P^2)} L$$

Assume measurements of R_D and R_P are collected from a dancer with $R_D = 5.4$ cm and $R_P = 9.1$ cm. Substituting these values into the above expression gives

$$z_{CM} = 0.417L$$

meaning that the longitudinal CM position is 41.7 % of length from the hip joint center to the knee joint center. The CM position of the thigh is closer to the hip than the knee, as expected given the larger proximal diameter of the segment.

One advantage of using eq. (3.2) to find the CM position of individual body segments, like the above example, is that it is based on direct body measurements of an individual. However, it still relies on the assumptions that distribution of mass $\rho(\vec{x}) = $ constant and the choice of a simplified shape for the segment.

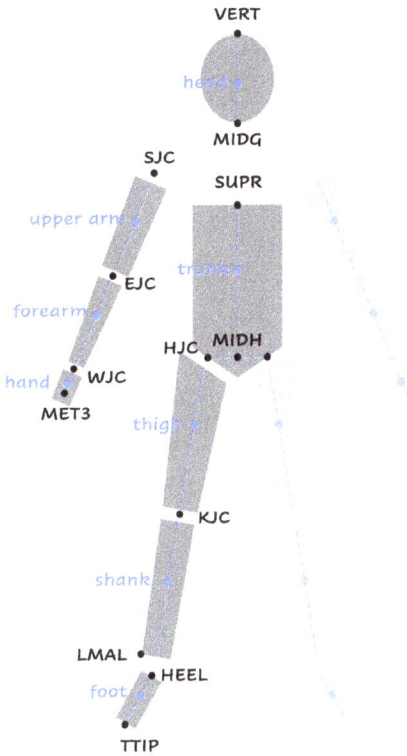

Figure 3.5: Body segments used for a segmental analysis. Individual segment CM positions (blue) can be treated as discrete point particles to find the whole body CM position. Segment origins and endpoints (black) are VERT = most superior point on the head; MIDG = midpoint between the gonions (jaw angle), SUPR = most superior point on the sternum jugular notch, MIDH = midpoint between hip joint centers, SJC = shoulder joint center, EJC = elbow joint center, WJC = wrist joint center, MET3 = third metacarpale, HJC = hip joint center, KJC = knee joint center, LMAL = lateral malleolus, HEEL = posterior point of the heel, TTIP = tip of the longest toe.

An alternative and widely accepted method to estimate an individual segment's CM position is through using previously published anthropometric data. *Anthropometric data* are compiled from measurements of the body's size and shape, providing information on average body segment parameters for populations such as segment masses, lengths, or CM positions. Researchers have experimentally measured body segment properties via cadaver studies (e. g., [16]) and more modern imaging methods like MRI or gamma scanning (e. g., [116]). The cadaver studies primarily considered older white males, but more recent studies are somewhat more inclusive in terms of age and gender. Ideally, anthropometric data would be available for a diversity of bodies. A widely utilized source for anthropometric data on young adults is that of Zatsiorsky et al. [116], which was then adjusted by de Leva [20] to use joint centers (vs. bony landmarks) as reference points (Fig. 3.5). The data are summarized in Tables 3.1 and 3.2. (Note that

Table 3.1: Body segment masses as a percentage of total body mass (as reported in [20]). The trunk may be represented as a single segment or three separate segments (upper, middle, and lower trunk). Definitions of segment origins and endpoints are given in Fig. 3.5 and XYPH = xyphion or substernale, OMPH = omphalion or center of the navel.

Segment	Origin	Endpoint	Mass, Female (%)	Mass, Male (%)
Head	VERT	MIDG	6.68	6.94
Trunk	SUPR	MIDH	42.57	43.46
Upper Trunk	SUPR	XYPH	15.45	15.96
Middle Trunk	XYPH	OMPH	14.65	16.33
Lower Trunk	OMPH	MIDH	12.47	11.17
Upper Arm	SJC	EJC	2.55	2.71
Forearm	EJC	WJC	1.38	1.62
Hand	WJC	MET3	0.56	0.61
Thigh	HJC	KJC	14.78	14.16
Shank	KJC	LMAL	4.81	4.33
Foot	HEEL	TTIP	1.29	1.37

Table 3.2: Body segment longitudinal center of mass positions (as reported in [20]). CM positions are measured from the segment Origin and reported as a percentage of the segment length. Definitions of segment origins and endpoints are given in Fig. 3.5 and XYPH = xyphion or substernale, OMPH = omphalion or center of the navel.

Segment	Origin	Endpoint	CM Position, Female (%)	CM Position, Male (%)
Head	VERT	MIDG	58.94	59.76
Trunk	SUPR	MIDH	41.51	44.86
Upper Trunk	SUPR	XYPH	20.77	29.99
Middle Trunk	XYPH	OMPH	45.12	45.02
Lower Trunk	OMPH	MIDH	49.20	61.15
Upper Arm	SJC	EJC	57.54	57.72
Forearm	EJC	WJC	45.59	45.74
Hand	WJC	MET3	74.74	79.00
Thigh	HJC	KJC	36.12	40.95
Shank	KJC	LMAL	44.16	44.59
Foot	HEEL	TTIP	40.14	44.15

our model of the thigh as a frustum of a cone gives a similar result as the average CM position for males in Tab. 3.2.)

One advantage of using anthropometric data is convenience. Individual segment masses can be estimated directly from total body mass, easily measurable with a scale. For example, if a female dancer's total mass is $M = 70.0$ kg, her upper arm mass is approximately 2.55 % of M (Tab. 3.1) or 1.79 kg. Individual segment CM positions can be determined relative to the positions of segment endpoints. If these segment endpoints are tracked via motion capture or a video analysis of a dance movement, the process of computing total body center of mass position can be automated frame by frame. A notable disadvantage of using anthropometric data is that it represents averages over populations and will not reflect individual variability.

Example. Use the anthropometric data in Tables 3.1 and 3.2 to estimate the CM position of the dancer in Fig. 3.6.

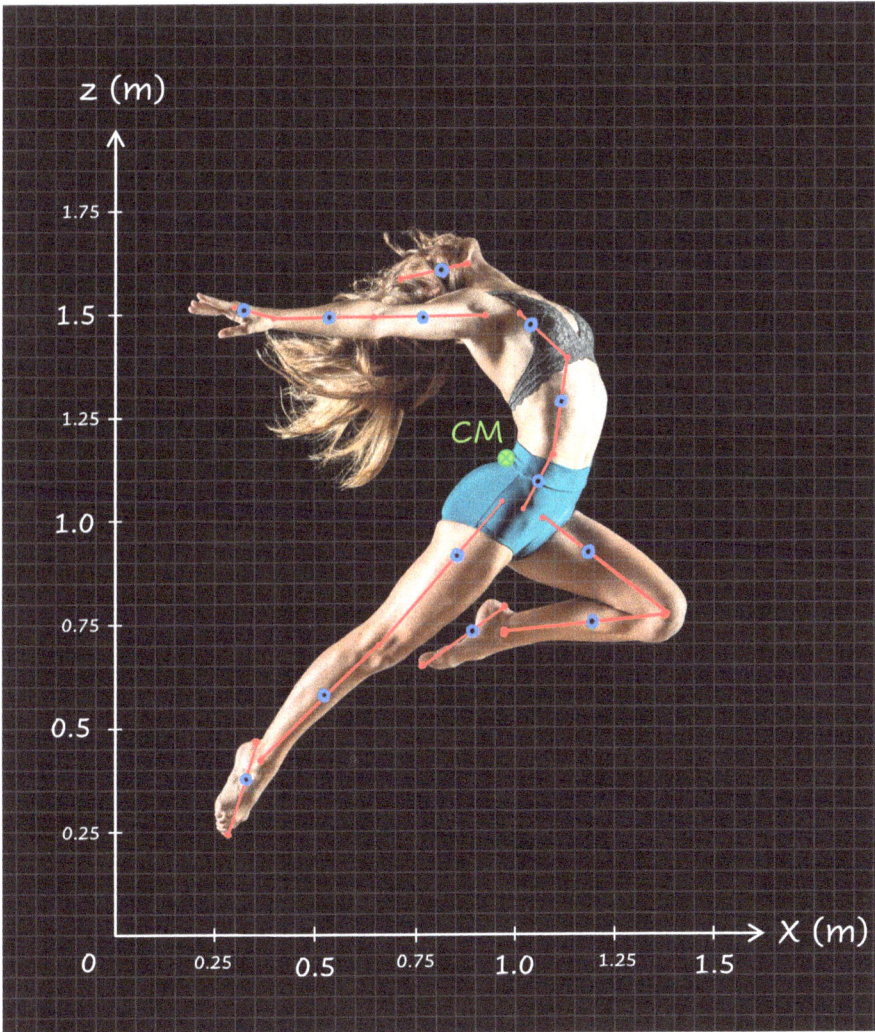

Figure 3.6: Segmental analysis to determine a dancer's whole body CM position. Body segment endpoints (red) are used to locate segment CM positions (blue data points), then compute the whole body CM position (green data point) with eq. (3.1). Photo by David Hofmann on Unsplash.

Solution. A line was first drawn between the endpoints of each segment (projected onto the longitudinal axis) using the landmarks in Tab. 3.2 (e. g., between the hip joint center (HJC) and knee joint center (KJC) for the thigh). The three segment representation of the trunk was chosen (upper, middle, and lower trunk) due to the back extension in this pose. Each segment's length was measured, then the segment's CM longitudinal position was computed using the percent distance from the segment origin in Tab. 3.2. A data point was placed at the segment's CM position, and its x- and z-coordinates were recorded using the grid provided. Since the left arm is blocked from view, it's position is assumed to be equal to that of the right arm. Segment CM coordinates for all 16 body segments are displayed in Tab. 3.3. Whole body x_{CM} and z_{CM} were computed from these data using eq. (3.1) and the segment fractional masses in Tab. 3.1. For the horizontal position,

$$x_{CM} = \sum_{i=1}^{16} \frac{m_i}{M} x_i = \left(\frac{m_{head}}{M} \right) x_{head} + \cdots + \left(\frac{m_{Lfoot}}{M} \right) x_{Lfoot}$$

$$= (0.0668)(0.82\,\text{m}) + \cdots + (0.0129)(0.89\,\text{m})$$

$$= 0.97\,\text{m}.$$

And for the vertical position,

$$z_{CM} = \sum_{i=1}^{16} \frac{m_i}{M} z_i = \left(\frac{m_{head}}{M} \right) z_{head} + \cdots + \left(\frac{m_{Lfoot}}{M} \right) z_{Lfoot}$$

$$= (0.0668)(1.61\,\text{m}) + \cdots + (0.0129)(0.73\,\text{m})$$

$$= 1.14\,\text{m}$$

Table 3.3: Segment CM positions in the xz-plane for the dancer in Fig. 3.6.

Segment	x (m)	z (m)
Head	0.82	1.61
Upper Trunk	1.04	1.47
Middle Trunk	1.12	1.28
Lower Trunk	1.06	1.09
R Upper Arm	0.77	1.49
L Upper Arm	0.77	1.49
R Forearm	0.53	1.49
L Forearm	0.53	1.49
R Hand	0.32	1.51
L Hand	0.32	1.51
R Thigh	0.85	0.91
L Thigh	1.18	0.92
R Shank	0.52	0.58
L Shank	1.19	0.75
R Foot	0.33	0.38
L Foot	0.89	0.73

The whole body CM position was found to be $\vec{x}_{CM} = (x_{CM}, z_{CM}) = (0.97\,\text{m}, 1.14\,\text{m})$. We observe that the dancer's CM is more posterior than its location in anatomical position and is nearly outside of the body. Note that it is possible for the CM to be located outside of the body depending on the relative positions of body segments.

As shown in the above example, the process of locating segment CM positions and calculating whole body CM position by hand is quite laborious, even for a single still photograph. In reality, we can streamline the process of tracking the whole body CM position throughout time as a dancer moves by performing a frame-by-frame video analysis of the motion. With digitized video data, tracking of segment endpoints and computing CM positions can be automated with a computer.

3.2 Center of mass kinematics

Now that we've discussed the physical meaning of center of mass and how to determine the a dancer's CM position from the positions of body segments, we move on to describe how the CM moves through space and time. To do so, we first choose an appropriate reference frame and coordinate system. In Chapter 2, our preferred reference frame was attached to the body (*body frame*), and we chose a *local coordinate system* (mediolateral, anteroposterior, longitudinal axes with their origin at the CM) that remained fixed *relative to the dancer's body*. So, if the dancer's CM accelerates, the body frame accelerates with it. In this chapter, we will select a reference frame fixed to the unmoving dance floor or stage (*stage frame*), and a *global coordinate system*, for example, a Cartesian coordinate system with the origin at center stage, the x-axis pointing toward stage right, the y-axis pointing downstage (toward the audience), and the z-axis vertically up. Unlike the body frame, the stage frame does not move or accelerate.

In Cartesian coordinates, the dancer's *position* is specified by the x, y, and z coordinates of the CM in the stage frame (x, y, z), or written in component notation:

$$\vec{x}_{CM} = x\hat{x} + y\hat{y} + z\hat{z} \tag{3.3}$$

where $\hat{x}$, $\hat{y}$, and $\hat{z}$ are unit vectors in the x, y, and z directions, respectively. The distance of the CM from the origin at any time is $\|\vec{x}_{CM}\| = \sqrt{x^2 + y^2 + z^2}$. The *displacement* of the CM from some initial position $\vec{x}_{CM,0}$ to a final position $\vec{x}_{CM}$ is defined as $\Delta\vec{x} = \vec{x}_{CM} - \vec{x}_{CM,0} = (x - x_0)\hat{x} + (y - y_0)\hat{y} + (z - z_0)\hat{z}$.

Instantaneous *velocity* is the time rate of change of position

$$\vec{v}_{CM} = \frac{d}{dt}\vec{x}_{CM} = \dot{\vec{x}}_{CM} \tag{3.4}$$

where the dot notation is used to represent the time derivative. Velocity can also be written in component form:

$$\vec{v}_{CM} = \frac{dx}{dt}\hat{x} + \frac{dy}{dt}\hat{y} + \frac{dz}{dt}\hat{z} = \dot{x}\hat{x} + \dot{y}\hat{y} + \dot{z}\hat{z}$$

Instantaneous *acceleration* is the time rate of change of velocity

$$\vec{a}_{CM} = \frac{d}{dt}\vec{v}_{CM} = \frac{d^2}{dt^2}\vec{x}_{CM} = \ddot{\vec{x}}_{CM} = \ddot{x}\hat{x} + \ddot{y}\hat{y} + \ddot{z}\hat{z} \tag{3.5}$$

where the double dot notation represents the second time derivative. Graphically, instantaneous velocity is the slope of the tangent line of position vs. time curve, and instantaneous acceleration is the slope of the tangent to a velocity vs. time curve. Acceleration also often changes with time during human movement. The time rate of change of acceleration (third time derivative of position) is referred to as "jerk." However, jerk and other higher-order time derivatives of position are less commonly computed in practice.

Example. Fig. 3.7 is an illustration of the *y*-component position vs. time (measured in centimeters (cm)) for a dancer doing a forward lean. The CM begins at the origin, then moves downstage (*y*-direction) as the dancer leans forward. Then the CM position decreases back to $y = 0$ as the CM moves back to its original position in one smooth, continuous motion. Use the *y* vs. *t* plot to sketch the velocity (*ẏ* vs. *t*) and acceleration (*ÿ* vs. *t*) plots. At what times are the velocity and acceleration at maxima, minima, and zero?

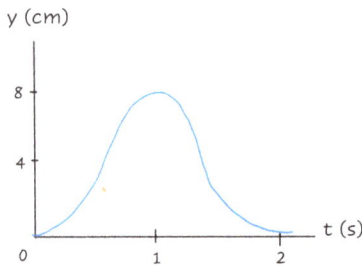

Figure 3.7: Sketch of CM position vs. time during a dancer's forward lean.

Solution. From $t = 0$ to $1\,\mathrm{s}$, the slope of *y* vs. *t* is positive, so the CM has a positive velocity (Fig. 3.8). The tangent to the curve begins with an approximately zero slope at $t = 0$, then its slope increases to a maximum value shortly after around 0.5 seconds. The slope decreases, and when the CM reaches its maximum position ($t = 1\,\mathrm{s}$), the velocity is momentarily zero. From $t = 1$ to 2 seconds, the CM moves back toward the origin, so the velocity is negative. The slope becomes more negative until around 1.5 seconds, at which point the slope becomes less and less negative, until the CM comes to rest again at $t = 2$ seconds.

A similar analysis is applied to determine the shape of acceleration vs. time from velocity vs. time (Fig. 3.8). Times at which the slope of the tangent to the *v* vs. *t* curve is zero indicates times in which acceleration is zero (at around $t = 0.6$ and $1.3\,\mathrm{s}$). When the

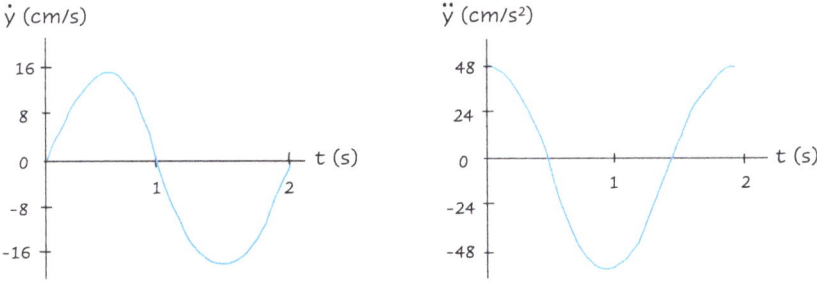

Figure 3.8: Sketches of CM velocity ($\dot{y}$) and acceleration ($\ddot{y}$) vs. time during a dancer's forward lean.

slope of v vs. t is positive, acceleration is positive; when the slope of v vs. t is negative, the acceleration is negative.

When kinematic data are measured experimentally (e. g., from analyzing video or motion capture data), the CM position can be tracked frame by frame in the video. Instead of a mathematical function or a smooth, continuous curve like the previous example, the data are digitized giving CM position at each discrete time point (Fig. 3.9). The time between data points is determined by the capture rate of the cameras. For example, a 60 Hz capture rate (60 frames per second) means the time between data points is 1/60 s. To compute CM velocity and acceleration from digitized position data, one option is to fit all or part of the position data with a curve (if it can be approximated by a continuous, analytical function), then finding $\vec{v}(t)$ and $\vec{a}(t)$ from its derivatives. Another common experimental practice is to use a *finite differences* method to determine velocities and accelerations. For example the velocity at a time instant (t_i) can be estimated by computing the average velocity between the data points immediately before (at t_{i-1}) and immediately after (t_{i+1}) that time (Fig. 3.9). For the x-component of velocity,

$$\dot{x}_i = \frac{x_{i+1} - x_{i-1}}{2\Delta t} \tag{3.6}$$

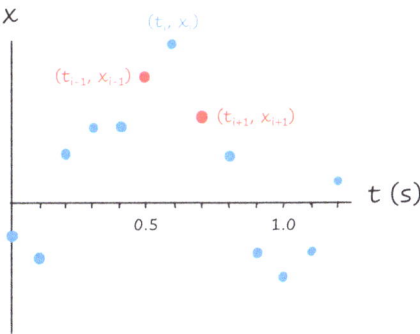

Figure 3.9: Digitized position vs. time data (capture rate = 10 Hz).

where Δt is the time between adjacent data points, with the same process being used to find the y- and z-components of velocity from their position components. The acceleration can be computed in a similar manner from the digitized velocity data.

$$\ddot{x}_i = \frac{\dot{x}_{i+1} - \dot{x}_{i-1}}{2\Delta t} \tag{3.7}$$

Experimental measures of body landmarks (and therefore total body CM position) have uncertainty due to the manner in which data are collected. If segment endpoints are tracked using markers placed on the dancer's skin or clothing, some of that uncertainty comes in the placement of the markers (e. g., whether the measured hip joint center is at the true hip joint center). Markers are placed superficially, whereas we wish to define segment endpoints using landmarks beneath the skin. During movement, the skin and soft tissues move relative to the bones and markers can experience vibrations (e. g., when large accelerations occur, like during jump landings), resulting in high-frequency noise in the data.

Using finite differences techniques tends to amplify the high frequency noise in the data, so acceleration results can be rather noisy. There are various techniques that can be used to smooth the data (e. g., digital filtering). For more information on smoothing experimental kinematic data, see for example Robertson, et al., *Research Methods in Biomechanics*, 2nd ed. [83].

Example. CM position data of a dancer doing a forward lean were experimentally collected (capture rate $=$ 120 Hz). The position data were low-passed filtered with a second-order Butterworth filter (cutoff frequency $=$ 3.5 Hz). The first several data points are given in Tab. 3.4 and all data are displayed graphically in Fig. 3.10. Comparing to the sketch in the earlier example, we can see that the data follow a similar overall shape. Use finite differences to compute and plot CM velocity ($\dot{y}$) and acceleration ($\ddot{y}$) vs. time.

Table 3.4: Experimental data measuring a dancer's CM position during a forward lean at a capture rate of 120 Hz.

t_i	t(s)	y(cm)
t_0	0.0000	−0.126
t_1	0.0083	−0.108
t_2	0.0167	−0.089
t_3	0.0250	−0.068
t_4	0.0333	−0.045
⋮	⋮	⋮

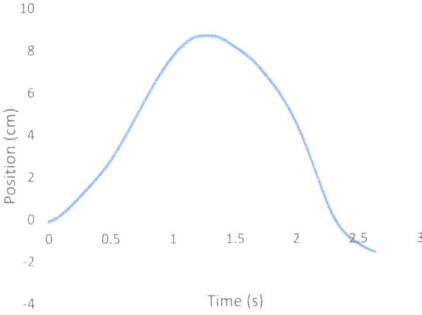

Figure 3.10: Experimentally measured CM position (y) of a dancer doing a forward lean. Although the graph appears continuous due to the high capture rate (120 Hz), position data are in fact discrete data points.

Solution. Using eq. (3.6) to compute velocity, we observe that our first velocity data point will occur at t_1. We cannot compute a velocity at t_0 because there is no measurement of y_{-1}. From the given data,

$$\dot{y}_1 = \frac{y_2 - y_0}{2\Delta t} = \frac{(-0.089 - (-0.126))\,\text{cm}}{2(0.00833\,\text{s})} = 2.221\,\text{cm/s}$$

The process is repeated to compute $\dot{y}_2$, $\dot{y}_3$ and so on (Tab. 3.5). The last velocity data point will occur one time frame prior to the final position data point. The loss of one initial and one final time frame once again occurs when acceleration data are computed. For the first acceleration data point, eq. (3.7) gives

$$\ddot{y}_2 = \frac{\dot{y}_3 - \dot{y}_1}{2\Delta t} = \frac{(2.64 - 2.22)\,\text{cm/s}}{2(0.00833\,\text{s})} = 25.2\,\text{cm/s}^2$$

This process is repeated for all available frames and the results are displayed graphically in Fig. 3.11.

Table 3.5: Velocity and acceleration are computed using the finite difference method from digitized experimental measures of CM position.

t_i	t (s)	y (cm)	$\dot{y}$ (cm/s)	$\ddot{y}$ (cm/s^2)
t_0	0.0000	−0.126		
t_1	0.0083	−0.108	2.22	
t_2	0.0167	−0.089	2.40	25.2
t_3	0.0250	−0.068	2.64	28.8
t_4	0.0333	−0.045	2.88	27.7
⋮	⋮	⋮	⋮	⋮

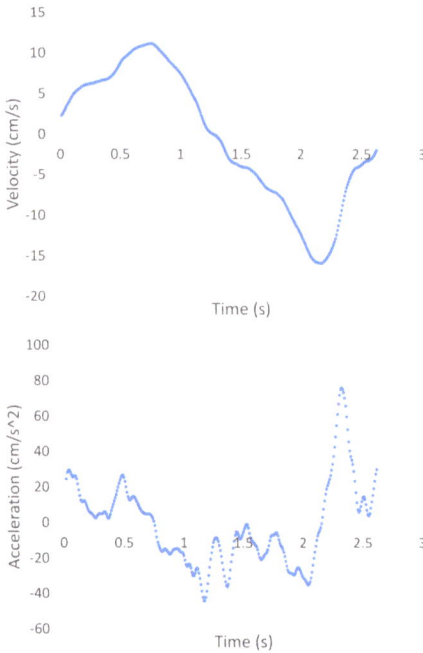

Figure 3.11: CM velocity and acceleration of a forward lean, computed from experimentally measured position data using finite difference method.

The overall shape of the velocity curve follows that predicted in the previous example, but the experimentally collected data are not as smooth. The velocity initially increases in the positive direction as the dancer leans forward, then slows down (becomes less positive) as she reaches her maximum forward leaning position at 1.3 s. At this point, velocity is momentarily zero as her CM changes direction. Her CM speed increases in the negative direction then slows down once again as her body reaches her final, upright standing position. Comparing the acceleration plot to the previously sketched prediction, we see that the experimental data have higher frequency oscillations on top of the overall trend that acceleration starts out as positive, becomes negative for the middle portion, then returns to a positive acceleration at the end of the motion. The high-frequency oscillations are primarily due to the data capturing actual human movement, including some jerky movements to maintain balance during the lean. There is also likely random uncertainty in the data that remains after filtering.

Suppose instead we had measures of CM acceleration with time and we wished to compute velocity and position as functions of time from the data. We would not have direct measures of acceleration from video data, but could obtain it instead from measurements of external force or with an accelerometer. Again, we begin with the definition of acceleration, $a = dv/dt \longrightarrow dv = a\,dt$, and integrate both sides to obtain $\Delta v = v - v_0 = \int a\,dt$ or

$$v = v_0 + \int a\, dt$$

Knowing acceleration as a function of time $a(t)$ and the initial velocity v_0, we could predict $v(t)$ by performing the integration from the initial to final time. Graphically, this integral is represented by the area under the curve of $a(t)$. Again, we do not often know the functional form of $a(t)$ during dance, but if we have digitized data, we can perform a digitized integral. If the capture rate is high, so the time between adjacent acceleration data points is small, we can make the assumption that acceleration is constant for this small Δt and to find the velocity in the frame $i = 1$, $\Delta v = a\Delta t$ becomes

$$v_1 = v_0 + a_0 \Delta t$$

and the process iterates to find the velocity at any time frame i,

$$v_i = v_{i-1} + a_{i-1}\Delta t$$

Similarly, to find $x(t)$ we start with the definition of velocity, $v = dx/dt \longrightarrow dx = v\, dt$ and integrate to find $\Delta x = x - x_0 = \int v\, dt$ or

$$x = x_0 + \int v\, dt$$

or for digitized data, again assume acceleration is constant for a very small Δt and find x_i in each time frame,

$$x_i = x_{i-1} + v_{i-1}\Delta t + \frac{1}{2}a_{i-1}(\Delta t)^2$$

3.3 Center of mass dynamics: Forces and Newton's laws

In the previous section, we described how a dancer moves without considering *why* they move the way that they do. In this section, we discuss the forces that cause CM motion and how a dancer utilizes these forces to dance.

Students of introductory physics should recall Newton's three laws, which form the foundation of Newtonian mechanics:

1. When zero net force acts on an object, it maintains a constant velocity. One possible constant velocity is zero; hence, an object at rest remains at rest unless acted upon by a net nonzero force. The first law is also known as the "Law of Inertia."
2. An object's acceleration is directly proportional to the net force acting on the object and inversely proportional to its mass, $\vec{a} = \frac{\sum \vec{F}}{m} \longrightarrow \sum \vec{F} = m\vec{a}$. From this, we see that Newton's first law is just a special case of the second law.
3. The force that object 1 exerts on object 2 is equal in magnitude but opposite in direction to the force exerted by object 2 on object 1 ($\vec{F}_{1\,on\,2} = -\vec{F}_{2\,on\,1}$). These forces are often termed "action–reaction pairs."

An important note about the first and second laws is that they are in fact not valid in all reference frames. Imagine standing in a moving subway car that stops abruptly as it reaches a station. You are not holding onto anything for support, your body lurches forward, and you awkwardly bump into an unsuspecting passenger. All of the seated passengers observed your forward acceleration (and your resulting embarrassment). There was no forward force pushing you, so Newton's second law seems to have been violated. However, Newton's first and second laws are only valid in non-accelerating, or "inertial", frames. The reference frame of the observers on the accelerating subway car is non-inertial. Newton's second law is not violated according to the people standing on the platform as they wait for the subway. These observers watch your body continue to move forward without enough backward force to stop you along with the car. The distinction between inertial and non-inertial frames is important when we consider reference frames affixed to a moving dancer's body. For now, we will stick with inertial reference frames attached to a stationary ground.

A dancer's total body movement can be separated into motion of the CM and the motion of the body's various segments relative to the CM. When we focus on CM motion, we will see that Newton's second law can be applied to separately analyze the movement of the CM. A brief proof is presented here to validate this claim. We begin with the definition of the CM of our system (eq. (3.1)), where the sum is over the very large total number of particles making up the entire body (system), then we take the second time derivative of each side of the equation:

$$\vec{x}_{CM} = \frac{\sum_i m_i \vec{x}_i}{M} \longrightarrow \ddot{\vec{x}}_{CM} = \frac{\sum_i m_i \ddot{\vec{x}}_i}{M}$$

By Newton's second law, the mass times acceleration of the ith particle equals the sum of all of the forces on that particle $\sum_j \vec{F}_{j,i} = m_i \ddot{\vec{x}}_i$, so

$$M\ddot{\vec{x}}_{CM} = \sum_i \sum_j \vec{F}_{j,i}$$

Note that $\sum_i$ is the sum over all of the particles and $\sum_j$ is the sum of all of the forces on the ith particle. These forces are from sources both external and internal to the system. For example, the first particle may exert a force $\vec{F}_{1 \, on \, 2}$ on particle 2 (an internal force since it is due to another part of the system). Keep in mind that the second particle exerts an equal but opposite force on the first particle by Newton's third law ($\vec{F}_{2 \, on \, 1} = -\vec{F}_{1 \, on \, 2}$), so these internal forces cancel out when the double sum is evaluated. The result is that only forces from sources *external* to the system contribute, and we have Newton's second law for CM motion:

$$M\ddot{\vec{x}}_{CM} = \sum \vec{F}_{ext} \tag{3.8}$$

where M is the total mass of the system (dancer), $\ddot{\vec{x}}_{CM}$ is the CM acceleration, and $\sum \vec{F}_{ext}$ is the vector sum of the external forces (net external force) acting on the dancer. This result shows that indeed the dancer's CM moves like a point particle of mass M subject to the net external force on the dancer.

Let's interpret this result some more in the context of a dancer. The CM acceleration is determined only by the sum of external forces applied to the dancer, not forces internal to the system. Recall in this case the system is the dancer's entire body, so the internal forces include all of those forces from sources inside of the body, such as muscle forces or joint contact forces. These internal forces when summed will cancel, as, for example, the force that a muscle exerts on the bone to which it is attached is equal but opposite to the force that the bone exerts on the muscle. External forces are forces that are applied to the system from objects or fields outside of the system. The single substantive field force applied to the dancer is the force of gravity. The other external forces on a dancer are contact forces, so when determining what forces to include in our model, we simply ask the question, what is in contact with the dancer? The ground? A partner? A prop? A dancer is also in constant contact with the air, which may exert a resistive force (drag). These forces are discussed in more detail individually in the following sections.

3.3.1 Force of gravity

The *force of gravity* (F_g) is an attractive force between objects due to their masses. It is a field force, meaning that the massive objects need not be in contact for the force to act. At the surface of the Earth, the magnitude of the force of gravity that the Earth exerts on a point particle with mass m is

$$F_g = mg \qquad (3.9)$$

where $g = 9.81\,\text{m/s}^2$, and F_g acts vertically downward. The force of gravity is also often referred to as the object's *weight*. For a dancer with a constant mass at or near the Earth's surface, F_g is a constant value and always acts on the dancer whether they are on the ground, in the air, dancing in the water, etc. When the object or system is a collection of point particles or a continuous mass distribution like a dancer, we represent the force of gravity on the entire system as acting at the CM. If the dancing were instead on the Moon, Mars, or a different planet, we would simply replace the acceleration due to gravity with its appropriate value (e. g., $g_{Moon} = 1.6\,\text{m/s}^2$ and $g_{Mars} = 3.7\,\text{m/s}^2$).

Example. A ballet dancer performs a pas de chat – a traveling leap in which both feet are off of the ground, and the dancer moves in the horizontal and vertical directions. Video data are collected at 100 Hz and body segment CM positions are tracked frame by frame. Segmental analysis is performed to determine the dancer's CM position as a discrete function of time. From the data on vertical (z) and horizontal (x) positions with time, experimentally determine the CM vertical and horizontal accelerations during the traveling leap.

Solution. Ignoring air resistance, the only external force on the dancer while in the air is the force of gravity, so $\sum F_z = M\ddot{z} \longrightarrow -Mg = M\ddot{z}$, so the vertical acceleration is expected to be $\ddot{z} = -g$. Since no horizontal force acts on the dancer, we expect the horizontal acceleration $\ddot{x} = 0$. The dancer's CM will follow a parabolic trajectory, subject to kinematics equations for constant acceleration:

$$z = z_0 + \dot{z}_0 t - \frac{1}{2}gt^2 \qquad x = x_0 + \dot{x}_0 t$$
$$\dot{z} = \dot{z}_0 - gt \qquad \dot{x} = \dot{x}_0$$
$$\ddot{z} = -g \qquad \ddot{x} = 0$$

The following are plots of experimentally measured x vs. t and z vs. t (Fig. 3.12). The horizontal position data are fit with a linear regression and the line of best fit is displayed. Since the slope of x vs. t does not change, the horizontal velocity is constant ($\dot{x} = -0.799$ m/s) and horizontal acceleration is zero. The z vs. t data are fit with a polynomial, showing that the CM indeed follows a parabolic trajectory. Additionally, finite differences method (eq. (3.6)) was used to compute vertical velocity data points and the plot of these data are shown in Fig. 3.12. The velocity graph shows a constant negative acceleration and the line of best fit gives an experimental value of $\ddot{z} = -9.84$ m/s^2, which is very close to the expected value of $-g = -9.81$ m/s^2. Note that the statistical uncertainty in the slope was measured to be ± 0.04 m/s^2, so the measured value is equal to the accepted value within experimental uncertainty.

Let us now think back to the dancer from the opening of the chapter, who wants to stay in the air for a longer time during a leap. Once the dancer's feet leave the ground, no matter what she does with her body while in the air, the CM will follow a parabolic trajectory predetermined by the initial takeoff velocity ($\dot{x}_0$, $\dot{z}_0$). The amount of time spent in the air is determined only by the initial vertical velocity and the total vertical displacement of the CM during flight. Due to aesthetic and/or technique constraints, the dancer is unlikely to be permitted to land the jump with the CM in a lower position (e. g., how a long jumper lands in a nearly seated position). (Solving our kinematics equation for z in the case that the CM begins and ends its flight at the same vertical position ($\Delta z = 0$) results in the total time in the air $t = \frac{2\dot{z}_0}{g}$.) So, to spend more time in the air, the dancer must increase her vertical takeoff velocity ($\dot{z}_0$).

If CM trajectory can't be changed once the dancer's feet leave the ground, why is it that some dancers appear to defy gravity and float while in the air? This "floating illusion" has been observed in dance leaps such as ballet grand jeté, and the effect was quantified in a study [58]. Researchers found that a dancer's head moved horizontally (within about 2 cm) for more than half of the time spent in the air, whereas the CM followed its parabolic trajectory, traveling 10 cm vertically during that same time. This floating illusion can enhance the aesthetic effect of the leap. How is it that the CM can follow a parabolic path while the head travels along a nearly horizontal line? The answer has to do with the timing of the arms and legs while in the air. In the grand jeté

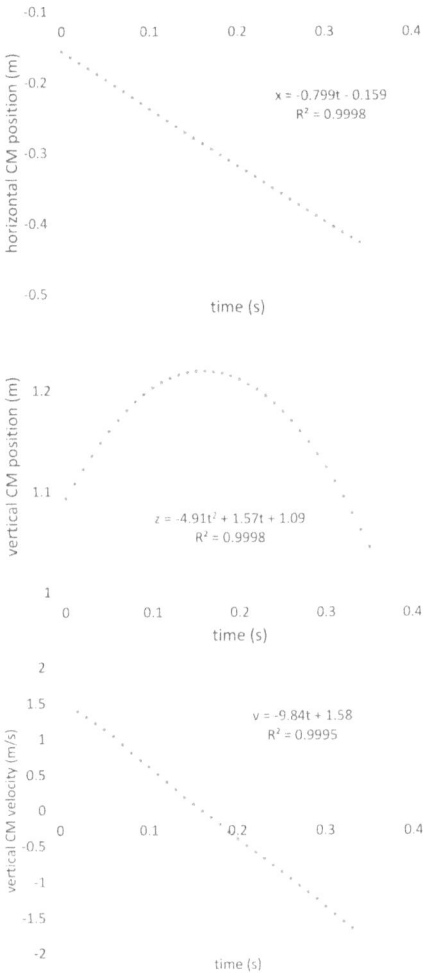

Figure 3.12: Horizontal CM position (top), vertical CM position (middle), and vertical CM velocity (bottom) vs. time for a dancer doing a 2D leap.

(Fig. 3.13), the legs and arms are lifted as the body approaches the peak of the jump. Then the arms and legs are lowered during the jump's descent. As we learned earlier, moving the arms and legs upward cause the body's CM to move superiorly (toward the head), then lowering them causes the CM to move inferiorly (back toward the feet).

While the dancer appears to float due to the head and torso moving horizontally, the amount of time a dancer spends in the air is not actually extended with the floating illusion. So what is a dancer to do in order to achieve the vertical take-off velocity necessary to stay in the air longer? To answer this question, we need to consider the push-off for the jump and the force that the ground exerts on the dancer to propel them into the air.

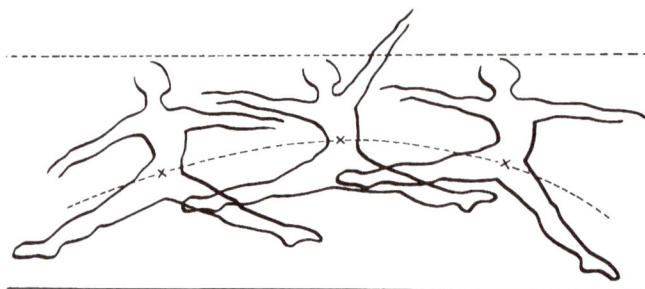

Figure 3.13: The dancer's CM (marked with an X) follows a parabolic trajectory, while the head follows a nearly horizontal path achieving the grand jeté "floating illusion." When the legs and arms are lifted, the CM moves superiorly within the dancer's body. Reproduced from K. Laws, "The Physics of Dance," Phys. Today 38, 24 (1985), doi:10.1063/1.880998, with the permission of American Association of Physics Teachers.

3.3.2 Ground reaction force

If a dancer's feet or any other part of the body is in contact with the ground, the ground exerts a force on the dancer. Biomechanists call this force the *ground reaction force* (*GRF*) because its action–reaction pair is the force that the dancer exerts on the ground (Fig. 3.14).

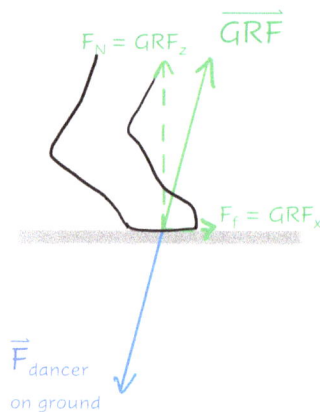

Figure 3.14: A dancer's foot pushing against the ground and the ground pushing back on the dancer. The ground reaction force $\overrightarrow{GRF}$ is the force that the ground exerts on the dancer and is equal but opposite to the force that the dancer exerts on the ground ($\overrightarrow{GRF} = -\vec{F}_{\text{Dancer on Ground}}$). The component of GRF perpendicular to the ground is the normal force (F_N), and the component parallel to the ground is the force of friction (F_f).

Unlike the force of gravity, which remains constant, both the magnitude and direction of the GRF can change dramatically, because it changes in response to the force that the dancer applies. If a standing dancer activates the muscles in the lower limbs, causing the legs to bend, the downward force that the dancer exerts on the ground is reduced and therefore the upward force of the ground on the dancer also decreases. When a dancer initiates forward movement from a standing position, the feet push backward against the ground, and in response, the ground pushes forward against the dancer, resulting in a forward acceleration of the CM.

Ground reaction forces are vital to accelerate the CM while dancing, but especially large or repeated forces from the ground that are transmitted to bones, muscles, and connective tissues may lead to injury. One study found dancers with patellar tendinopathy (an overuse injury commonly known as "jumper's knee" resulting in pain in the patellar tendon) had greater peak vertical GRFs and braking GRFs during landings from jumps [23]. Techniques and/or equipment that reduce peak GRFs are important tactics for injury prevention in dance.

The 3D ground reaction force vector can be dynamically measured directly with a *force plate*. Similar to motion capture data, force plate data are collected at a set sampling frequency (e.g., 1000 Hz or every 0.001 s). The GRF vector can be separated into two orthogonal components: one component acting perpendicular to the surface and one acting parallel to the surface. These components of GRF are in fact two familiar forces studied in introductory physics: the normal force and force of friction.

3.3.2.1 Normal force

When you stand on a bathroom scale, it reads your weight, right? Wrong! The bathroom scale measures the normal force. The *normal force* (F_N) is a contact force that a surface exerts on an object and acts perpendicular (i.e., "normal") to the surface. If you bounce up and down on the scale or rest your arm on the sink while standing on it, you can change the reading on the scale while your weight ($F_g = mg$) remains unchanged.

Example. A dancer stands on a force plate on a horizontal floor, then moves their body by flexing and extend their knees, moving the arms, etc. Their body is only ever in contact with the force plate, not another part of the ground, a wall, a partner, etc. Derive an expression to determine the dancer's vertical acceleration ($\ddot{z}$) at any time from measurements of F_N.

Solution. The vertical external forces acting on the dancer consist of the downward force of gravity and upward normal force, as shown in Fig. 3.15. Applying Newton's second law to the vertical direction (z), we have

$$\sum F_z = F_N - F_g = M\ddot{z}$$
$$F_N - Mg = M\ddot{z}$$

Note that if $\ddot{z}$ is downward (negative), that means $F_N < F_g$, and if $\ddot{z}$ is upward (positive), that means $F_N > F_g$. If $\ddot{z} = 0$, then $F_N = F_g$. The expression $F_N - Mg = M\ddot{z}$ can be solved for F_N if measurements of acceleration were collected (e.g., from a video analysis or with an accelerometer),

$$F_N = M(\ddot{z} + g) \tag{3.10}$$

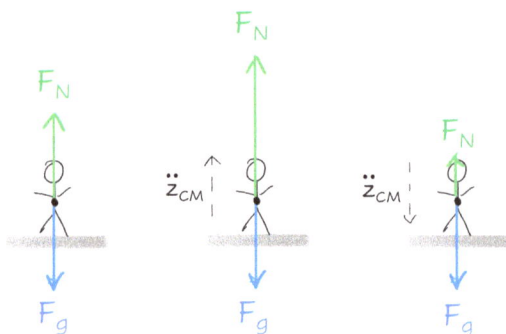

Figure 3.15: Vertical forces on a dancer in contact with the ground. When the CM acceleration ($\ddot{z}_{CM}$) is upward, $F_N > F_g$, and when $\ddot{z}_{CM}$ is downward, $F_N < F_g$.

or to find acceleration from normal force,

$$\ddot{z} = \frac{F_N}{M} - g.$$

Again thinking back to the dancer who wants her jump's air time to last longer, we see that a larger acceleration during pushoff is possible with a larger normal force. So we might think that if she just pushes harder downward against the ground, the ground will push up harder on her and that will do it. However, it is not quite this simple. We know that the change in velocity depends on both acceleration *and* the time over which the acceleration acts, and that F_N is in general non-constant. We will revisit this question a final time by analyzing actual data collected with a force plate for a dancer's pushoff for a jump later in the chapter.

When the ground is horizontal, F_N acts vertically upward. For this reason, biomechanists often refer to F_N as the vertical ground reaction force or vGRF. From eq. (3.10) we see that when $\ddot{z}$ is large and upward, F_N (vGRF) is also large. Large upward accelerations of the CM can occur during landings from jumps, as contact with the ground ceases the CM's downward motion. Researchers measured GRF during ballet saut de chat (a traveling leap from one leg to the other leg, similar to grand jeté), finding peak vGRF during landing to be ~4 times body weight [54]. Another study found that ballet assemblé (jump from one foot to two feet) recorded significantly lower total peak GRF (3.30 ± 0.44 times body weight) than grand jeté (3.77±0.91 times body weight) [70]. The grand jetés reached lower heights but covered much further horizontal distance than the assemblés, so they required large braking forces (horizontal GRF) during landing. Peak vGRF during landings from soft shoe Irish dance jumps (leap, birdie, and bicycle) have also been measured between ~3–4 times body weight [17].

Ground reaction force can spike when dancers perform percussive footwork, striking parts of the feet against the ground, as in tap dance or flamenco. Peak vGRFs during tap were measured to be comparatively lower (~2 times body weight) than other dance forms, possibly explaining the relatively lower prevalence of injuries in tap dancers [68]. Various footwork steps in flamenco recorded values similar to those found in tap, with the largest peak vGRF during tacón, or hitting the heel against the ground, approaching 3 times body weight [103]. Large peak vGRFs were recorded during hard shoe Irish dance stomps (averaged ~5 times body weight), with an individual peak force measured to be nearly 10 times body weight [17].

One final note about the normal force is that F_N is not necessarily vertical. Consider a dancer standing on a raked stage. The floor of a raked stage is inclined slightly such that it angles upward away from the audience. In this case F_N includes both horizontal and vertical components (Fig. 3.16). With the ground inclined at θ above the horizontal, the vertical component of the normal force is $F_{N,z} = F_N \cos \theta$ and the horizontal component is $F_{N,y} = F_N \sin \theta$.

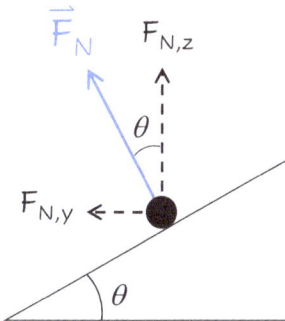

Figure 3.16: The normal force has vertical and horizontal components on an inclined surface like a raked stage.

3.3.2.2 Force of friction

For solo dancing where the body is only ever in contact with the horizontal ground, all center of mass motion comes down to how a dancer controls the vertical ground reaction force (F_N) and the horizontal ground reaction force (F_f). The sole reason a dancer's CM can accelerate horizontally is because of friction, since it is the only non-negligible external force in that direction. Friction is not only desirable but necessary; however, too much or too little friction can have negative consequences. When a dancer's foot needs to smoothly slide against the ground, too much friction can jolt the foot to a sudden stop. On the other hand, if a dancer must make a quick direction change with the feet planted firmly on the ground, too little friction can cause the dancer's feet to slide out from underneath the body, and the dancer could strain a muscle or fall. Not only is the appropriate amount of friction needed for peak performance, it is necessary to prevent injuries.

The *force of friction* (F_f) between two solid surfaces (like a dancer's foot and the floor) resists their relative sliding motion (or impending motion) and acts parallel to the surfaces in contact. Friction is categorized into two cases: *static friction*, in which there is no relative motion between the surfaces, and *kinetic friction*, in which the surfaces slide relative to one another. Dancers often develop an acute awareness about friction and the feeling of their feet or other parts of their body against the ground because the amount of friction greatly impacts their dancing.

If a breaker does a head slide with the head sliding to the right relative to the ground, the ground exerts a force of kinetic friction in the opposite direction (to the left) on the head (Fig. 3.17). The force of friction depends on the roughness of the surfaces in contact, characterized by an experimentally determined quantity, called the coefficient of friction (μ), and how hard the surfaces push against one another (the normal force, F_N). For kinetic friction,

$$F_{f,k} = \mu_k F_N.$$ (3.11)

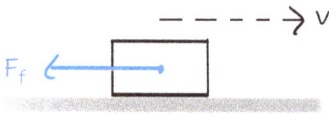

Figure 3.17: The force of kinetic friction F_f acting in the direction opposite to an object's velocity (v) as it slides against the ground.

For the static friction case, let's imagine a dancer pushes against the ground to take a step forward (Fig. 3.14) with no sliding between the foot and the ground. Imagine first that he exerts a small backward force on the ground, so the forward force of static friction on him and his horizontal acceleration are also small. To accelerate more rapidly forward requires a greater force (e. g., to begin running), so the dancer must push back harder on the floor, increasing the force of static friction. There is a point at which the static frictional force reaches its maximum value, so if he pushes too hard backward against the ground, his foot may slip, and he could possibly fall. The maximum force of static friction, like kinetic friction, depends on surface roughness and normal force ($F_{s,max} = \mu_s F_N$). Since the force of static friction increases until it reaches its maximum value, the force of static friction is

$$F_{f,s} \leq \mu_s F_N.$$ (3.12)

The coefficients of friction for several surfaces are given in Tab. 3.6. Rubber on concrete (e. g., walking in sneakers on the sidewalk) has a much greater coefficient of static friction than steel on ice (e. g., tap shoes on an icy sidewalk). Surfaces with larger coefficients of friction are rougher, less likely to result in slipping for static friction cases than smoother surfaces. On the other hand, decreasing kinetic friction is desirable in certain cases; for example, when the foot rotates against the ground during a turn. Not only is less friction advantageous when a dancer wants to keep spinning without slowing down, but decreasing friction can reduce undesirable joint torques transmitted to the knee as the foot twists against the ground.

Table 3.6: Coefficients of friction for various combinations of surface types. Several dance shoe sole surfaces (rubber, leather, metal) and flooring types (vinyl, wood) are displayed.

Surfaces	μ_s	μ_k
rubber on concrete	1.0	0.8
rubber on vinyl	0.2 to 0.3	–
leather on wood	0.3 to 0.4	–
metal on wood	0.2 to 0.6	–
wood on wood	0.25 to 0.5	0.2
rubber on ice	0.15	–
steel on ice	0.03	–
human synovial joints	0.01	0.003

Example. A dancer ($M = 64.0$ kg) stands at rest on a raked stage where the angle of incline is 6° above the horizontal. Find the dancer's weight, normal force, and force of friction acting on the dancer.

Solution. A free-body diagram (Fig. 3.18) provides a sketch of all of the external forces acting on the dancer.

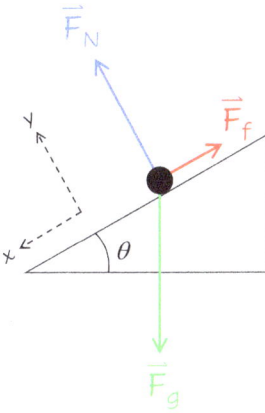

Figure 3.18: FBD of dancer standing at rest on a raked stage.

The dancer's weight is

$$F_g = Mg = (64\,\text{kg})(9.81\,\text{m/s}^2)$$
$$F_g = 628\,\text{N}$$

To find the other two forces, we can define our coordinate system such that the x-axis is parallel to the surface of the incline (aligned with F_f) and the y-axis is perpendicular to the incline (aligned with F_N). Then we apply Newton's second law to the x- and y-directions separately, noting that since the dancer is at rest, $\ddot{x}_{\text{CM}} = 0$. With this choice of coordinate system, $\vec{F}_g$ has both x- and y-components, where $F_{g,x} = F_g \sin \theta$ and $F_{g,y} = F_g \cos \theta$.

$$\sum F_x = 0 \qquad \sum F_y = 0$$
$$F_{g,x} - F_f = 0 \qquad F_N - F_{g,y} = 0$$

So, the force of friction and normal force are

$$F_f = Mg \sin \theta \qquad F_N = Mg \cos \theta$$
$$= (628 \text{ N})(\sin 6°) \qquad = (628 \text{ N})(\cos 6°)$$
$$F_f = 65.6 \text{ N} \qquad F_N = 625 \text{ N}$$

Incline angles of raked stages tend to be small, as in this example, so we notice that the normal force differs only slightly from its value on a horizontal stage (in this case, only a 0.5 % reduction). However, a non-negligible force of friction is necessary to keep a dancer's feet from sliding down the incline toward the audience. Using eq. (3.12), we find $\mu_s \geq 0.105$. The incline does change how much additional force a dancer can exert parallel to the incline before their feet begin to slip. The implication is that a dancer must modify how they accelerate the body. Moving down the incline is aided by gravity, but acceleration up the incline is made more difficult. Muscle activation patterns will need to be altered slightly to produce the same movements on the raked stage as compared to a horizontal stage.

Very few experimental studies have quantified coefficients of friction for different dancewear-flooring combinations. One study on country swing dance measured μ_s between a force plate and three shoe types: rubber bottom boots, leather bottom boots, and rubber bottom running shoes, finding that both types of rubber bottom shoes ($\mu_s \sim 0.6$ to 0.75) had higher coefficients of static friction than the leather bottom boots ($\mu_s \sim 0.2$ to 0.3) against the force plate [75]. While having a database of measured coefficients specific to different dancewear-flooring combinations (including bare skin) may be useful to help give a range of values for various dance styles, these coefficients vary depending on many factors. For example, the coefficient of friction between a canvas ballet slipper with a leather sole and vinyl flooring depends on which part of the shoe is in contact with the ground, how much the shoe has been worn, the type of vinyl flooring, the wear of the floor, etc. The floor may also have a variable roughness depending on which areas get more wear, if there are seams or tape on the floor, etc. The coefficient of friction between human skin and the floor depends on factors such as which part of the body (e. g., bare feet, hands, arms, etc.) is in contact with it, and how much sweat, hair, etc. cover it. If the coefficient of friction is required for a research study, it is best for it to be measured directly.

Dancers employ various methods to modify the force of friction between their feet (or other parts of the body) and the floor. The means leveraged by dancers most often have to do with changing the properties of the surfaces in contact (μ). Some dancers apply rosin (a type of resin) to their footwear (e. g., pointe shoes) to increase friction and prevent slipping. There are other dancers (e. g., contemporary or jazz dancers) who wish to decrease friction and might apply talcum powder to their bare feet or wear socks while dancing. Dancers must use caution with talcom powder, since too much can make the floor surface hazardously slippery. Patches of slick flooring left by talcom powder can be unexpectedly encountered by dancers, and for this reason its use is not allowed in some dance spaces. It is also important to have consistency between the flooring and footwear used in a rehearsal space and that used in a performance space. Unexpected differences in flooring (e. g., spending hundreds of hours rehearsing on a vinyl dance floor then dancing the same piece on a wooden stage in a performance venue) can have negative consequences for both a dancer's performance and their risk of injury.

The force of friction also changes if the normal force changes, so dancers can dynamically adjust F_f while dancing via vertical acceleration of the CM. This effect has been observed experimentally in dancers performing spins, or pirouettes [63]. Some of the dancers in the study did a subtle "lifting" motion of the body, and when their CM experienced a slight negative acceleration at the end of the lift, the dancers could slide their feet more easily against the ground.

3.3.3 Other external forces

3.3.3.1 Contact forces: Partners, props, walls
Not all dancing is solo dancing without props. In partnered dance, contact forces between another dancer's body are often present. A partner can exert a push or a pull on a dancer, whether by holding hands, standing back-to-back, or many other means of contact. Props exert forces on dancers, whether it is an object that is held by a dancer or a chair on which the dancer sits. Like the ground, the props can exert normal forces and forces of friction on the dancer via the surfaces in contact. Also like the ground, a vertical surface (e. g., a wall) can exert force on a dancer. In this case, the normal force would act horizontally, pushing outward from the wall onto the dancer. The force of friction would act parallel to the wall's surface (in the vertical direction). A rope or a cord could be used as a prop and exert a tension force (which can only pull) on a dancer.

Example. Two ice dancers with equal mass ($m_1 = m_2 = 65\,\text{kg}$) hold hands and spin. Their bodies move on a circular path (radius $= 75\,\text{cm}$) with their hands held at the center (top down view shown). Find the force that each dancer must exert on the other to remain spinning if their tangential speed is $v = 2.5\,\text{m/s}$.

Solution. The ice dancers move in uniform circular motion, in which Newton's second law in the centripetal direction is $\sum F_c = ma_c \longrightarrow \sum F_c = \frac{mv^2}{r}$, where v is the tangential speed and r is the radius of the circular path (Fig. 3.19). Since the dancers are on the ice, we will assume the force of friction is negligibly small, so the only force that acts in the

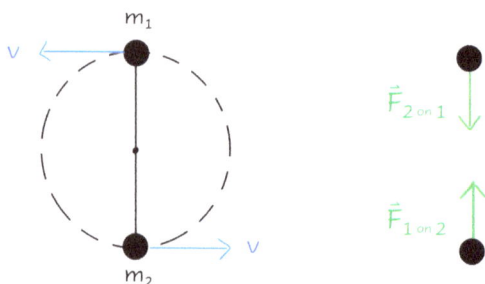

Figure 3.19: Two ice dancers holding hands and moving in uniform circular motion. The force that each dancer exerts on the other is equal but opposite by Newton's third law.

centripetal direction is the force that each dancer exerts on the other. To find the force that Dancer 2 exerts on Dancer 1, we apply $\sum F_c = \frac{mv^2}{r}$ to m_1.

$$\sum F_c = \frac{m_1 v^2}{r}$$

$$F_{2\,on\,1} = \frac{(65\,\text{kg})(2.5\,\text{m/s})^2}{(0.75\,\text{m})} = 542\,\text{N}$$

We obtain the same result for the force on Dancer 2, so each dancer would need to exert this rather large force ($542\,\text{N} \approx 122\,\text{lb}$) on the other to remain spinning. To put the tangential speed of 2.5 m/s into perspective, let's find the number of revolutions per second for the dancers ($\omega = \frac{v}{r} = 3.33\,\text{rad/s} = 0.53\,\text{rev/s}$). So the dancers are spinning at a rate at which it would take almost 2 seconds to make one complete revolution. To spin faster (and/or with a smaller radius) would require even greater centripetal force. *What if?* What if the dancers did not have equal masses (e. g., $m_1 = 65\,\text{kg}$ and $m_2 = 80\,\text{kg}$)? The dancers would still exert forces equal in magnitude on one another ($\vec{F}_{1\,on\,2} = -\vec{F}_{2\,on\,1}$ by Newton's third law). Since the dancer's arms are the same length, the same total distance of 1.5 m remains between their bodies. We realize that since $m_1 \neq m_2$, the dancers can no longer move at the same tangential speed with the same radius of curvature.

$$|F_{2\,on\,1}| = |F_{1\,on\,2}|$$

$$\frac{m_1 v_1^2}{r_1} = \frac{m_2 v_2^2}{r_2}$$

If the dancers are to always remain on opposite sides from one another, they must have the same rotational speed, ω, and since $v = \omega r$,

$$\frac{m_1 (\omega r_1)^2}{r_1} = \frac{m_2 (\omega r_2)^2}{r_2}$$

$$m_1 r_1 = m_2 r_2$$

So, the ratio of the radii of curvature of the two dancer's paths is $\frac{r_1}{r_2} = \frac{m_2}{m_1}$. This means that the pivot point must shift from the center of the circle to closer to the more massive dancer (Fig. 3.20).

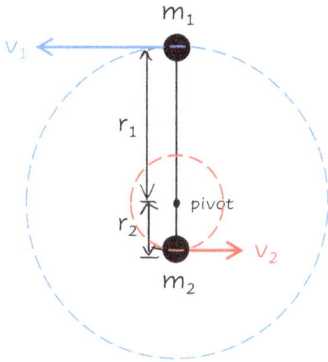

Figure 3.20: Ice dancers of unequal mass $(m_2 > m_1)$ holding hands and moving along circular paths.

With $m_1 = 65\,$kg and $m_2 = 80\,$kg, $\frac{r_1}{r_2} = \frac{80}{65} = 1.231$, and since $r_1 + r_2 = 1.5\,$m, $r_1 = 0.83\,$m and $r_2 = 0.67\,$m. If we again assume that ($\omega = 3.33\,$rad/s), then

$$F_{2\,\text{on}\,1} = \frac{m_1(\omega r_1)^2}{r_1}$$

$$= \frac{(65\,\text{kg})((3.33\,\text{rad/s})(0.83\,\text{m}))^2}{(0.83\,\text{m})}$$

$$= 598\,\text{N}$$

It makes sense that a somewhat larger force is required in this scenario, since Dancer 1 (having the same mass as before) must now complete the same number of revolutions per second, but on a path with a larger radius of curvature. Note also that the pivot point is at the CM position of the two-dancer system, which remains stationary since the net external force on the two-dancer system is zero.

3.3.3.2 Air resistance

When a dancer moves through the air (or any other fluid), the air exerts a resistive force on the body. Two primary sources of fluid resistance are 1) surface drag (surface friction as the fluid drags along surface of the object) and 2) pressure drag (collisional forces exerted by the fluid particles as they strike the object). For slow-moving objects in viscous fluids, surface drag tends to dominate, whereas for human movement in less viscous air, pressure drag dominates as the air molecules must be pushed out of the way to allow the body to move through.

Anyone who has been outside on a windy day has experienced the increased force that air exerts on the body as the body's speed relative to the air (v) increases. The pressure *drag force* ($\vec{F}_D$) acts in a direction opposite to the velocity of the object relative to the fluid ($-\hat{v}$), and its magnitude is found from

$$F_D = \frac{1}{2}\rho A C_D v^2$$

where ρ is the density of the fluid (ρ_{air} = 1.29 kg/m^3), A is the cross-sectional area of the object perpendicular to $\hat{v}$, and C_D is the coefficient of drag, an experimentally deter-mined, unitless quantity that quantifies how aerodynamic (or not) the object is. For a human body, $C_D \approx 1.0$, and $A \approx 0.4$ m^2 for the frontal plane (largest cross sectional area). So $F_D \approx 0.26v^2$ for the drag force during dancing. A world-class sprinter reaches a maxi-mum speed of $\approx$ 10 m/s during the 100-meter dash; most dancing occurs at much slower speeds. A plot of drag force as a function of body speed relative to the air is shown in Fig. 3.21.

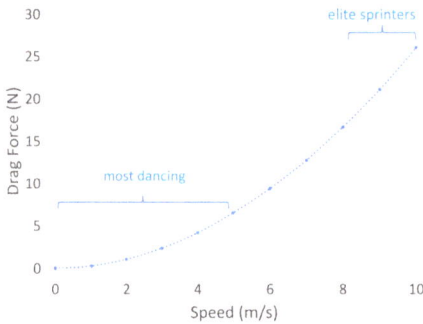

Figure 3.21: Force of air resistance (drag) on the human body as a function of the body's speed (relative to the air).

Since the drag force tends to be negligibly small compared to other forces acting on a dancer (GRF and F_g), in many cases it is justifiable to ignore drag for biomechanical analyses of dancing. Examples of cases when drag may need to be considered are when large pieces of fabric are worn or held (increasing cross-sectional area like a parachute), if dancing occurs outside on a windy day (so v reaches larger values than dancing in still air), or if the dancer is in water (ρ_{H_2O} = 1000 kg/m$^3 \gg \rho_{air}$).

3.4 Impulse-momentum theorem

Other than the force of gravity, the external forces acting on a moving dancer in general vary with time, resulting in a non-constant CM acceleration. Instead of using Newton's second law directly to determine a dancer's motion as a function of time, we can recast it into an alternate mathematical form known as the impulse-momentum theorem. As we will see, the theorem is particularly useful if we wish to know the overall change in momentum of a dancer (and therefore the dancer's total change in velocity) given how the net force on the dancer varies with time. One such instance is the beginning of chapter example, in which a dancer wished to remain in the air for a longer time during a leap. By measuring GRF during push-off, we can determine the dancer's takeoff velocity the moment their feet leave the ground.

In a few simple steps we can obtain the impulse-momentum theorem beginning with Newton's second law.

$$\sum \vec{F} = m\vec{a} = m\frac{d\vec{v}}{dt}$$

Since the dancer's mass m is constant, and using the definition of *linear momentum*, $\vec{p} = m\vec{v}$, we have

$$\sum \vec{F} = \frac{d(m\vec{v})}{dt}$$
$$= \frac{d\vec{p}}{dt}$$

Then multiplying both sides by dt and integrating,

$$\sum \vec{F}\, dt = d\vec{p} \longrightarrow \int \sum \vec{F}\, dt = \int d\vec{p}$$

$$\int_{t_0}^{t_f} \sum \vec{F}\, dt = \Delta\vec{p} \tag{3.13}$$

where the integral is from the initial time, t_0, to the final time, t_f and $\Delta\vec{p} = \vec{p}_f - \vec{p}_0$ is the change in momentum over that time interval. We define the quantity on the left hand side of eq. (3.13) as the *impulse*, $\vec{J} = \int \sum \vec{F}\, dt$, hence eq. (3.13) is termed the *impulse-momentum theorem*. Whereas Newton's second law in its original form reads "A net force causes a mass to accelerate," the impulse-momentum theorem reads, "A net impulse causes a system to change momentum." Impulse is a 3D vector ($\vec{J} = J_x\hat{x} + J_y\hat{y} + J_z\hat{z}$), so we can find the x-, y-, and z-components of impulse from the x-, y-, and z-components of net force, respectively:

$$J_x = \int \sum F_x\, dt, \quad J_y = \int \sum F_y\, dt, \quad J_z = \int \sum F_z\, dt \tag{3.14}$$

Graphically, impulse is the area under the curve of net force vs. time (Fig. 3.22). Greater forces exerted over a longer duration of time result in larger impulses and therefore larger change in momentum. A large force may result in only a small change in momentum if exerted over a very small duration of time. During a dancer's push-off from the floor for a 2D traveling leap, $\sum F_x = \pm F_f$ in the horizontal direction, and $\sum F_z = F_N - F_g$

vertically. (The same holds true when the feet regain contact with the ground during the landing from a jump, in both cases, the force of friction can act as either a propulsive $(+F_f)$ or braking force $(-F_f)$). To find the change in horizontal and vertical components of velocity during push-off (or landing) a jump, eq. (3.13) gives

$$\Delta v_x = \frac{1}{m} \int F_f \, dt$$

$$\Delta v_z = \frac{1}{m} \int (F_N - F_g) \, dt$$

To increase the vertical takeoff velocity, and therefore time spent in the air, our dancer must increase the vertical impulse during push-off. This is achieved by increasing F_N and/or the time duration of the active push-off phase. F_N is an external force; however, its magnitude depends on how much force the dancer exerts on the ground, which is in turn determined by forces produced by muscles inside of the body. So not only do we need the impulse-momentum theorem, but an understanding of how the muscles work to explain how a dancer can jump higher. The following example analyzes real, experimental data to illustrate the effects of different push-off techniques.

ΣF (N)

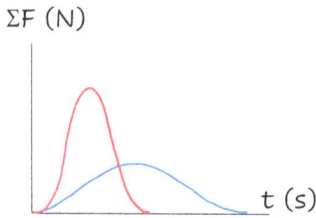

Figure 3.22: Impulse is the area under the net force vs. time curve. The red force vs. time has a greater peak force, but is exerted over a smaller amount of time compared to the blue force. Each have approximately equal areas under the curve (shaded regions), so the impulses, and therefore changes in momentum, are approximately the same.

t (s)

Example. A dancer ($m = 61.1\,\mathrm{kg}$) pushes off for a vertical jump from ballet first position (feet flat on the ground with the heels together and toes pointing outward). The dancer jumps as high as possible using two different push-off methods. In the first case (A), the dancer begins at rest in plié (knees flexed). After holding the static plié for a few seconds, she pushes off as hard as she can for the jump, holding her arms down at her sides (Fig. 3.23). In the second case (B), she begins with the legs straight and arms out. Then, in one fluid motion, she bends and then straightens the legs, moving the arms down and then up along with the legs to push off for the jump (Fig. 3.24). In both cases, she uses the same amount of knee flexion (same depth of plié). A force plate measures F_N (vGRF) at a frequency of 1000 Hz during the entire jumping motion (push-off, free flight (in air), and landing) (Figs. 3.25 and 3.26). Use the force data to determine the dancer's takeoff velocity, time spent in the air, and maximum jump height for each case.

Solution. We begin by interpreting the vGRF plots. For the jump beginning from a static plié, vGRF $= F_g$ then increases to a local maximum of 1568 N as the dancer pushes against the ground and her CM experiences an upward acceleration (Fig. 3.25). Then

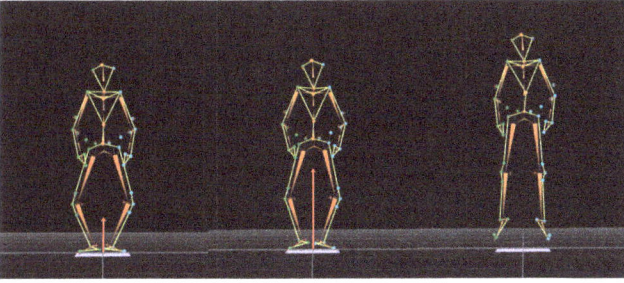

Figure 3.23: Case A: Jump from static plié position. (Red arrow is the GRF vector.)

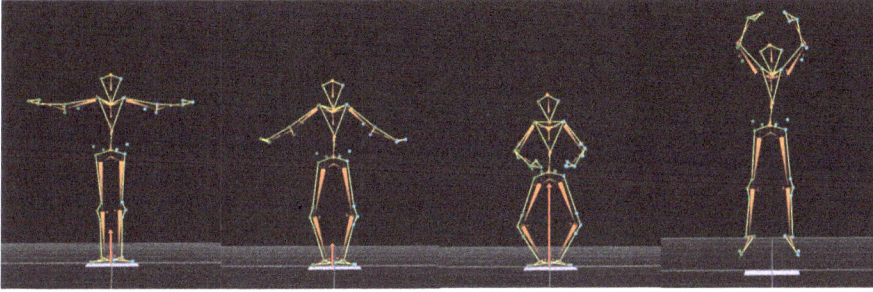

Figure 3.24: Case B: Jump with a countermovement. Red arrow is the GRF vector.

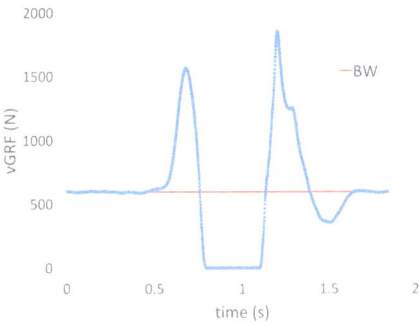

Figure 3.25: Case A: Vertical ground reaction force for dancer doing a jump starting from a static plié position. Solid line is dancer's body weight (F_g = 599 N).

vGRF reaches zero when her feet leave the ground at the start of the free-flight phase. When her feet again make contact with the force plate for the landing of the jump, vGRF increases sharply, then decreases as the dancer bends the knees to cushion the landing.

For the jump with a countermovement, vGRF first decreases from body weight then increases to a local maximum of 1295 N during the push-off (Fig. 3.26). vGRF $< F_g$ when the knees begin to flex and the CM has a downward acceleration. When the CM begins to accelerate back upward, vGRF $> F_g$ before vGRF returns to zero for the free flight phase. Once again, vGRF peaks shortly after landing, then decreases and eventually levels back out at the dancer's body weight when she stands at rest after the jump.

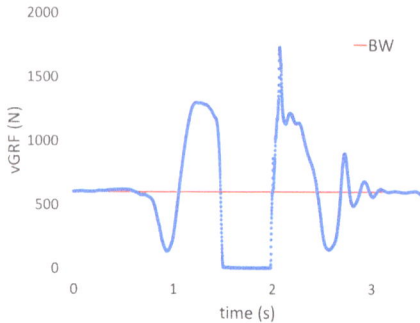

Figure 3.26: Case B: Vertical ground reaction force for dancer doing a jump with a countermovement. Solid line is dancer's body weight (F_g = 599 N). The dancer's landing was more unstable for Case B and is why vGRF oscillates more before leveling out than the plot in Case A.

To solve for the dancer's takeoff velocity, we apply the impulse-momentum theorem to the push-off phase. Since the force plate measures vGRF at discrete time intervals, instead of performing an integral, we do a discrete sum to add up all of the infinitesimal impulses (dJ_i):

$$J_z = \sum dJ_i = \sum [(\text{vGRF})_i - F_g]\, dt$$

where dt is the sampling time interval of the force plate (0.001 s). The result for Case A is J_{zA} = 91.9 N s, and for Case B J_{zB} = 144.8 N s. In both cases, the dancer begins the push-off from rest. Once the vertical take-off velocity is determined from the impulse-momentum theorem, we find the time spent in the air (t), and maximum height of the CM measured from its standing position (z_{max}) for each case using kinematics. Results are compared below.

Case A: Static Plié

$$\text{Peak vGRF} = 1568\,\text{N}$$

$$J_z = 91.9\,\text{N s}$$

$$v_z = \frac{1}{m}J_z = 1.50\,\text{m/s}$$

$$t = \frac{2v_z}{g} = 0.307\,\text{s}$$

$$z_{max} = \frac{v_z^2}{2g} = 0.114\,\text{m}$$

Case B: Countermovement

$$\text{Peak vGRF} = 1295 \text{ N}$$

$$J_z = 144.8 \text{ N s}$$

$$v_z = \frac{1}{m}J_z = 2.40 \text{ m/s}$$

$$t = \frac{2v_z}{g} = 0.483 \text{ s}$$

$$z_{max} = \frac{v_z^2}{2g} = 0.294 \text{ m}$$

The dancer's CM went over 2.5 times higher and she spent more than 1.5 times the amount of time in the air when using the countermovement. This occurred even though the peak vGRF was actually *less* in Case B. Additionally, the negative part of the impulse (when vGRF < F_g and the graph dips below the red line in Fig. 3.22) detracts from the upward change in momentum, but the enhanced positive impulse more than makes up for this for an overall greater positive result.

So why is it that the jump with the countermovement allows for a greater total impulse during pushoff? Several factors, both physical and biomechanical, contribute. A physical factor has to do with extending the amount of time over which the propulsive force acts. The depth of plié is roughly the same in both cases; however, in Case B with the countermovement, the CM begins to accelerate back upward before the dancer reaches her maximum plié depth. So the amount of time with vGRF > F_g is extended. Another factor that allows the CM to accelerate upward for a longer time in Case B is the movement of the arms. As the arms move upward during push-off, the body's CM moves upward within the body, and this extends the amount of time possible for the feet to remain on the ground, continuing to exert force. Ken Laws (author of *Physics and the Art of Dance* [57]) coined the term "storing momentum" for this type of effect in which the movement of one part of the body stores momentum so that another part of the body (in this case the lower limbs) can have reduced momentum for a portion of time. The upward movement of the arms allows the legs to move upward less rapidly, and they can therefore continue to push against the ground.

A biomechanical factor contributing to impulse generation has to do with how the muscles are activated during push-off. Two primary joint actions when moving from the plié position to jumping are knee extension and hip extension (also ankle plantarflexion). In Case A, the dancer's knee extensor muscles (quadriceps group) and hip extensors (hamstrings and glutes) are activated isometrically in the starting plié position. Then the extensor muscles activate concentrically to execute the push-off. In Case B, however, the muscles are rapidly activated *eccentrically* immediately prior to their concentric activation. Let's consider the knee extensors (quadriceps). When the body moves downward, the knee extensors become active to slow the downward movement (providing upward CM acceleration). But as the body continues to move downward (with knee flexion), the quadriceps muscles continue to lengthen while they are activated. As learned in Chapter 2, rapid eccentric activation is capable of producing larger muscle force than isometric or concentric activation. The eccentric activation may increase the elastic potential energy storage in the muscle-tendon unit. If the stretching is very rapid, it may also elicit the stretch-reflex initiated by the muscle spindles, increasing the number of motor units recruited to concentrically activate the muscle. This rapid elongation of the activated muscles prior to their shortening is termed the *stretch shorten cycle* (SSC) and can enhance force production during movements like jumping and running.

The dancer, at the beginning of the chapter, equipped with all of this knowledge, should focus on their *vertical* take-off velocity (since horizontal velocity does not affect the time spend in the air), work on the timing of their arms to coordinate with the push-off (if possible given the choreography), and utilize a countermovement to take advantage of the SSC (again, if possible given the choreography). Plyometric training (e. g., drop jumps in which a dancer drops down from a raised surface and immediately and explosively jumps vertically upward using a rapid countermovement) may help the dancer develop the muscle strength and neuromuscular coordination necessary to help jump higher.

3.4.1 Reducing injury risk

As discussed previously, impacts with the ground can subject dancers to ground reaction forces multiple times larger than their body weight. Reducing the magnitude of these GRFs may help dancers avoid injury. When a dancer is moving with a downward velocity $\vec{v} = -v\hat{z}$ right before the feet hit the ground, it will take a set amount of upward impulse to stop the dancer's downward momentum ($\vec{J} = \Delta\vec{p} = 0 - (-mv)\hat{z} = mv\hat{z}$). We now know that increasing the time over which the GRF acts decreases the force required to produce the same impulse (Fig. 3.22). A deeper plié during the landing is one way to extend impact time. Landing with stiffer legs may increase the risk of injuries to the lower extremities; however, some leg stiffness is necessary to avoid soft tissue injuries such as an ankle sprain. One study found that dancers' leg stiffness was significantly less during saut de chat landings compared to push-offs as the dancers absorbed the impacts of the jumps [54]. Not all dance forms stylistically permit reduced leg stiffness upon landing, however. For example, Irish dance technique requires rigid landings with very little knee flexion or ankle dorsiflexion [17].

Dance footwear and flooring can also influence the peak vGRFs sustained while dancing. Similar to a deeper plié, footwear with more cushioning can slightly extend the time over which impact forces act. It may also help absorb some of the force so it is not transmitted directly to the foot. One study found no significant differences in peak GRF for dancers landing from jumps wearing pointe shoes, ballet flats, or bare feet [70], but both types of footwear provide relatively little padding for the feet. On the other hand, padded socks worn by modern and contemporary dancers were shown to significantly decrease vGRF and mean and peak heel forces (measured with circular force transducers on the dancer's feet within the socks) compared to jump landings in bare feet [85].

Sprung (suspended) floors, a type of lower stiffness dance flooring, may also help reduce injury risk. A flooring with less stiffness can help absorb impact forces and extend the amount of time over which GRF acts on a dancer, leading to lower peak GRFs. One study found that dancers landed grand jetés with less leg stiffness on un-sprung floors than on sprung floors [31], suggesting that the sprung floor required the dancers legs to absorb less impact force. Another study found that floors with lower force reduction capabilities (stiffer floors) increased the mechanical demands placed on dancers' ankle joints during landings, particularly during the early impact of the foot with the floor [35]. Variability in dance floor stiffness may also contribute to injury risk. One study found the highest injury rate in a professional ballet company was associated with rehearsing on the floor with greatest stiffness variability, not the overall stiffest floor; however, none of the floors in the study fell within the range of stiffness values advised by the European Sport Surface Standards [36].

Studying the external forces required for a dancer's center of mass motion has allowed us to understand how whole body movement is produced and how this understanding can be leveraged for performance enhancement and injury prevention. In the next chapter, we will focus on not one but two important points of the dancer's body – the CM and the dancer's point of support with the floor – to learn more about something that is on every dancer's mind: balance.

4 Balance

Two dancers joke around during a break between classes. One dancer, who happens to be standing on relevé, tells an especially funny joke, causing the friend to laugh and give the joking dancer a friendly shove. The dancer on relevé makes a graceful arm movement to remain balanced, impressing the friend. "How is it that you are always so good at keeping your balance?" the friend asks. "You hardly ever wobble during the hardest things we do in class, and you make it look so easy!" The dancer thinks for a minute and says, "I try to stay grounded, engage my core, and look at my spot." "I do that too!" laments the friend. "Really what's your secret?" The dancer things again and says truthfully, "I don't know," and after some more thinking says, "I mean, I practice on a balance disk at home, so I am sure that helps. When we're in class, I do the things I already mentioned and just feel for balance." The conversation leaves the dancer very curious. "Why *am* I so good at staying on balance?"

Dance often intersperses stationary poses or slow-moving, impressive balancing tasks throughout dynamic movement. Dancers can also transition smoothly and seamlessly between dynamic, multiplanar movements, often while balancing on a very small base of support. The ability to maintain balance in varied dynamic and static situations is crucial for dancers. With a strong emphasis on seemingly effortless performance of challenging balancing tasks, it is no wonder that dancers have developed a reputation for being balancing "experts." That reputation is backed by research, particularly as the level of dance training increases. Superior performance on balancing tasks has been measured for ballet dancers (e. g., [18] and [82]), Indian classical dancers [10], Thai classical dancers [52], and modern dancers [3] compared to non-dancers, for example.

A comprehensive look at the research shows that dancers exhibit enhanced performance compared to non-dancers in some, but not all, balancing tasks or conditions [3]. Differences have been found between dancers and non-dancers in the weighting of sensory inputs for balance maintenance (e. g., visual and somatosensory information) [93] and the different motor outputs and/or kinematics produced [87, 89]. The bottom line is that dancers seem to be different from non-dancers in how they sense for balance, how their muscles are activated, and how they move their bodies for balance maintenance.

In this chapter, we will study balance maintenance from physical and neuromuscular perspectives and dive deeper into the research on balance maintenance in dance. We will begin with the simplest physical model of a dancer balancing in a pose: the simple inverted pendulum. While there are clear limitations to this oversimplified model, including the fact that we will start with a passive model, useful information can still be gained from this simplified approach. We will then take into account how the body's mass is distributed and the inertia of different mass distributions and body configurations. Increasingly complex models will then be considered, taking into account greater relative motion between body segments and providing insight to dynamic body adjustments for balance maintenance. After discussing physical models and kinematic strate-

https://doi.org/10.1515/9783110642292-004

gies for balance maintenance, we will combine the roles of the nervous and musculoskeletal systems to discuss how dancers take in and integrate sensory information and produce the motor output necessary to maintain balance. Finally, we will discuss why dancers or dance training may lead to differences in balance performance from other populations. While there is still knowledge to be discovered about balance and dance, this chapter will help the curious balancing dancer at the beginning of the chapter more fully understand what it takes to be a balancing "expert."

4.1 Simple inverted pendulum model

A fixed pose may be held on one or two feet, the hands, knees, head, or any other parts of the body. In this chapter, we will consider dancers in balanced or unbalanced poses to only be supported via contact with the ground. Regardless of the complexity in body configuration or what parts of the body are used as the base of support, only two external forces act on the dancer: the force of gravity and the GRF. Not only do the magnitude and direction of these two forces matter, but the positions at which the vectors act. A small force of gravity acts on every small bit of mass making up a dancer's body, but the overall effect is the same as a single force of gravity equal to body weight acting at the body's center of mass (CM). The GRF is distributed over the entire surface of contact between the ground and the body, but the overall effect is the same as the total GRF acting at a single point called the center of pressure (CP). Fundamentally, balance maintenance comes down to the relative positioning of these two specific points. The simple inverted pendulum is our most basic physical model that includes both the CM and CP.

A mass hanging from a light string or rod is a physical system often modeled as a simple pendulum (Fig. 4.1). The simple pendulum is an idealized model with a single point mass M suspended from a massless rod of length L and a fixed pivot point. When displaced from its vertically hanging equilibrium position, the pendulum exhibits oscillatory motion due to the tangential force of gravity acting as a restoring force. If the displacement from equilibrium is small, the simple pendulum's motion is a classic example of simple harmonic motion. Because the direction of the tangential force of gravity is toward equilibrium, it is considered a stable equilibrium position.

A person in an upright or inverted posture can be modeled as an *inverted simple pendulum* (Fig. 4.2), or a point mass M (at the CM) on a rigid, massless rod of length L with a single point of support (at the CP). Unlike the regular simple pendulum, the inverted simple pendulum does not experience a restoring force if displaced from its equilibrium position (with the CM along a vertical line directly above the CP). Instead, gravity causes the pendulum to accelerate away from this unstable equilibrium position. Before using this model in the context of balance maintenance in dance, let's first discuss the meaning of center of pressure in more detail.

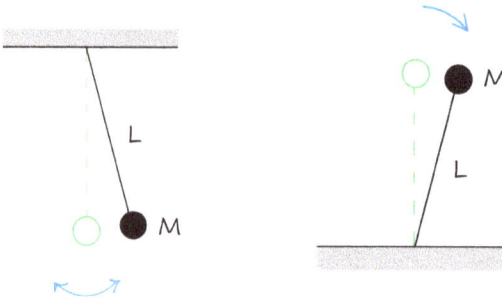

Figure 4.1: Simple pendulum model. Simple pendulum (left) with mass *M* suspended from a massless string of length *L* and the *inverted* simple pendulum (right) with mass *M* above its pivot point on a rigid, massless rod of length *L*.

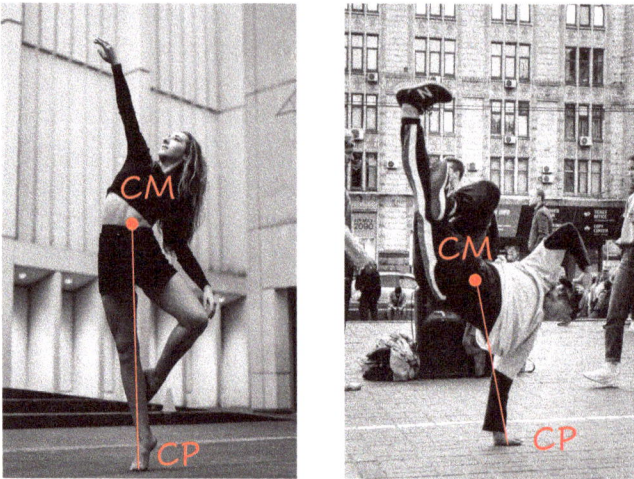

Figure 4.2: Dancer in an upright posture (left) and inverted posture (right), modeled as a simple inverted pendulum. Photos by Jeffery Erhunse on Unsplash (left) and Vicky Hladynets on Unsplash (right).

4.2 Center of pressure and base of support

The center of pressure is the point at which we consider the single GRF vector to act, so the CP is the same as the center of force application. Before we build up to a more general idea of CP, let's consider a simpler scenario. Imagine an object such as a beam with an unknown mass distribution placed on two points of support (Fig. 4.3). The normal force at the first support (position $\vec{x}_1$) is F_{N1} and the normal force at the second support (position $\vec{x}_2$) is F_{N2}. In general, F_{N1} does not equal F_{N2}. We define a center of force (CF) as a weighted average position of the normal force application

$$\vec{x}_{CF} = \frac{\vec{x}_1 F_{N1} + \vec{x}_2 F_{N2}}{F_{N1} + F_{N2}}$$

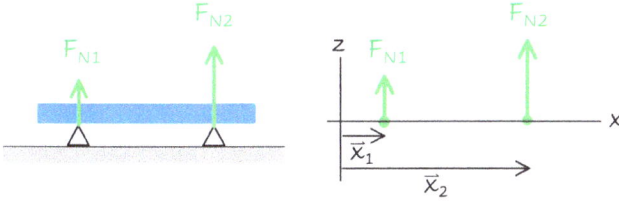

Figure 4.3: Beam with two points of support.

For any discrete number of points of force application

$$\vec{x}_{CF} = \frac{\sum \vec{x}_i F_{N,i}}{\sum F_{N,i}} \tag{4.1}$$

For a continuous distribution of force over a surface, like a foot in contact with the ground, we consider the infinitesimal portion of the GRF, $d\vec{F}$, exerted over the infinitesimal area dA (Fig. 4.4). $d\vec{F}$ has components perpendicular (dF_N) and parallel (dF_f) to the surface. In this case, the discrete sum in eq. (4.1) becomes an integral to find the center of normal force application.

$$\vec{x}_{CF} = \frac{\int \vec{x}\, dF_N}{\int dF_N}$$

Pressure is the perpendicular component of force per unit area over a surface ($P = \frac{F_\perp}{A}$), and for the GRF, $F_\perp = F_N$. We can then make the substitution that $dF_N = P(\vec{x})\, dA$. In general, the GRF is distributed over the base of support in a way that is non-uniform for human movement, which means that the pressure varies with position, $P(\vec{x})$.

$$\vec{x}_{CP} = \frac{\int \vec{x}\, dF_N}{\int dF_N} = \frac{\int \vec{x} P(\vec{x})\, dA}{\int P(\vec{x})\, dA}$$

$$= \frac{1}{F_N} \int \vec{x} P(\vec{x})\, dA \tag{4.2}$$

The *center of pressure* (CP) is the weighted average position of the normal force distributed over a surface, and eq. (4.2) provides the mathematical definition. We see that to compute CP from a pressure distribution is essentially the same as how we computed the CM from a mass distribution (eq. (3.2)). In biomechanics research, the CP position can be measured experimentally with a force plate (Fig. 4.5). In this case, researchers need not use eq. (4.2) since CP is a direct measurement output.

As with the CM, the CP can exist outside of an actual surface of contact. For example, imagine standing with the feet hip-width apart and equal weight supported by each foot. The CP will be at a point directly between the feet. Instead of imagining two separate normal force vectors acting at each foot, we can imagine one normal force vector acting at the CP. The CP is however constrained to reside within the boundaries of the *base of support*, which is not necessarily the same as the contact area.

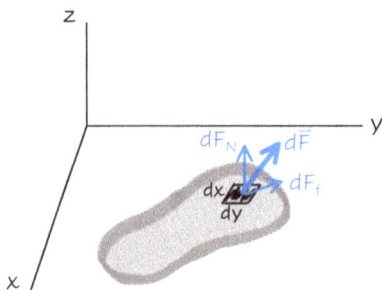

Figure 4.4: GRF exerted over an infinitesimally small portion of the contact area between the ground and foot.

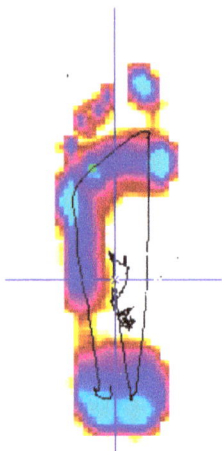

Figure 4.5: Experimentally measured pressure distribution. Center of pressure trajectory is in black. Researchers asked the participant to lean as far as possible in every direction while holding a support. Reprinted with permission from [34].

Consider for example, standing in two different positions: 1) with the feet together or 2) with the feet hip-width apart in parallel position (Fig. 4.6). Even though the same surface area of the bottom of the feet is in contact with the ground, position 2) has a wider base of support (BoS). The BoS can be found by considering an imaginary rubber band that stretches around the outside edges of the supporting body parts (Fig. 4.6). The CP can move within the BoS as more or less pressure is placed on different parts of the feet. You can feel greater pressure move from your heels to your toes if you shift your weight anteriorly. You can also move the CP to the left or right, and there is much more room for medial-lateral motion of the CP with the wider BoS. Standing in ballet first position (using turnout) versus the parallel position in case 1) does two things. It increases the medial-lateral range available for CP movement, but it decreases the anterior–posterior range. The concept of BoS is important when we discuss strategies for balance maintenance later in the chapter. The traditional criteria for balance maintenance is that the CM remain on a vertical line over the BoS, so the BoS area is important when considering stability. As we will see, dynamic balance maintenance is more complex than this criteria.

Figure 4.6: Base of support (BoS). BoS boundary shown in red standing with the feet in parallel first position (left), parallel second position (center), and ballet first position (right).

4.3 Simple inverted pendulum dynamics

For a dancer to remain perfectly in static equilibrium when balancing in a pose, they must remain at rest with exactly zero velocity and zero acceleration. No matter how we model the dancer, for a living, breathing human body, these conditions are unrealistic, even in a relatively straightforward pose like standing on two feet. For the net force and net torque to remain zero, the dancer's CM must never deviate, even a fraction of a millimeter, from directly vertical above the CP. Clearly this is realistically impossible, so let's first find what useful information may be obtained from considering the dynamics of toppling from equilibrium assuming the (passive) inverted pendulum model. In other words, what happens if the dancer topples holding the body rigidly, without making any adjustments to compensate for imbalance?

Since our simple inverted pendulum model considers the point of support to be fixed and the length of the pendulum to remain constant, the point mass (CM) is constrained to move on a circular path around an axis (the "topple axis") at the point of support. To describe the dynamics of toppling, we use the rotational analog of Newton's second law:

$$\sum \vec{\tau} = I\vec{\alpha} \tag{4.3}$$

where $\sum \vec{\tau}$ is the net external torque (in units of N·m), I is the moment of inertia (kg·m^2), and $\vec{\alpha} = \ddot{\vec{\theta}}$ is the angular acceleration (rad/s^2). The definition of *torque* exerted by a force $\vec{F}$ applied at a position $\vec{r}$ from an axis of rotation is given by the vector cross product:

$$\vec{\tau} = \vec{r} \times \vec{F} \tag{4.4}$$

The force $\vec{F}$ may have components parallel and perpendicular to $\vec{r}$, but it is the perpendicular component of $\vec{F}$ ($F_\perp$) that contributes to the torque. From the vector cross product, the magnitude of the torque can be found by

$$\tau = rF \sin \phi$$
$$= rF_\perp$$
$$= r_\perp F$$

where ϕ is the angle between $\vec{r}$ and $\vec{F}$. The quantity $r_\perp$ is known as the *moment arm* and provides us with another means to conceptualize the torque due to a force (Fig. 4.7). The moment arm is the component of $\vec{r}$ perpendicular to the line of action of the force $\vec{F}$. In the case of this simple inverted pendulum, the force of gravity contributes the only non-zero torque, because the GRF acts directly at the axis of rotation ($\vec{r}_{GRF} = 0$). *Moment of inertia* (I) depends both on an objects mass and how that mass is distributed around an axis. For the simple pendulum of length L, $I = ML^2$, so eq. (4.3) becomes

$$\sum \vec{\tau} = I\vec{\alpha}$$
$$\vec{\tau}_g = ML^2\ddot{\theta}$$
$$MgL\sin\theta = ML^2\ddot{\theta}$$

resulting in the equation of motion for the topple angle, θ:

$$\frac{g}{L}\sin\theta = \ddot{\theta} \tag{4.5}$$

As the topple angle increases, $r_\perp$ also increases, leading to a greater torque and therefore greater angular acceleration.

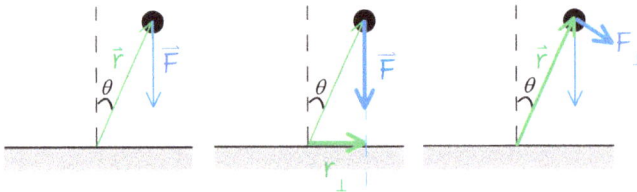

Figure 4.7: Torque due to the force of gravity around the topple axis of rotation at the CP. The moment arm ($r_\perp$) is the component of $\vec{r}$ perpendicular to the line of action of $\vec{F}$ (center). $F_\perp$ is the component of $\vec{F}$ perpendicular to $\vec{r}$ (right).

Before attempting to solve this non-linear differential equation to determine θ as a function of time, we first note some interesting physical results. First, the angular acceleration does not depend on mass, so our model predicts that dancers with different body masses but the same body stature will topple at the same rate. The two factors other than θ that affect angular acceleration are g (constant on the surface of the Earth) and the length of the pendulum, L. Since $\ddot{\theta}$ is *inversely* proportional to L, pendula that are longer have a lower angular acceleration than shorter pendula. In other words, if the CM is further from the ground (e. g., a taller dancer) the dancer will topple *less rapidly* than a lower CM.

This last point is worth highlighting since it may seem counterintuitive, particularly to dancers who may have been told in their dance training that lowering the CM leads to greater stability when dancing. That statement is simply not true! Keeping all other things equal (base of support area, body joint angles, etc.), lowering the CM leads to a less "stable" pose since a larger angular acceleration gives the body less time to make adjustments to correct for an imbalance. Comparing standing with the legs straight (and knees "locked") versus standing with the knees flexed in plié is not a fair comparison of relative stability. The knee flexion enables a dancer to more readily make necessary adjustments to correct for an imbalance and has nothing to do with the fact that the CM is lower in plié. The dancer at the beginning of the chapter mentioned that she "stays grounded" as a tactic for maintaining balance. That imagery can still be useful for dancers, even though it doesn't mean to move the CM closer to the ground. Perhaps it sends heightened awareness to the cutaneous receptors at the feet to sense CP movement or to other important proprioceptors in the feet and ankles.

A more appropriate comparison would be between standing with the arms held down versus the arms raised to the sides (90 degrees of shoulder abduction) or held above the head (180 degrees shoulder flexion). A dancer may find it challenging to balance on the ball of one foot with the arms down, but less challenging if their arms are lifted. Raising the arms effectively raises the whole body CM, and most importantly it increases the dancer's moment of inertia by placing more mass further from the topple axis of rotation. This is the same effect as a tightrope walker holding a long pole, increasing inertia and effectively increasing the resistance to angular acceleration.

Now let's get back to solving our differential equation (eq. (4.5)). Large topple angles (θ) can occur purposefully while dancing as part of choreography, for example a fall that is caught by a partner, falling forward before tucking the body into a forward roll, or while taking a large step in the direction of topple. Typically, however, for a quietly balanced pose, topple angles remain relatively small. Otherwise, dancers tend to take more drastic measures, like moving the base of support or changing it entirely, invalidating our fixed point of support assumption. Therefore it is useful to use the small angle approximation, so our differential equation can be solved analytically with elementary functions. When θ is "small," the series expansion of $\sin \theta = \theta - \frac{\theta^3}{3!} + \frac{\theta^5}{5!} - \frac{\theta^7}{7!} + \cdots \approx \theta$, since the higher order terms become vanishingly small. Then the equation of motion becomes

$$\frac{g}{L}\theta = \ddot{\theta} \quad \text{(for small topple angles)} \tag{4.6}$$

Those who have studied simple harmonic motion may notice that eq. (4.6) looks very similar to the equation of motion for a simple harmonic oscillator ($-\omega^2 x = \ddot{x}$) with one *very* important difference: a missing negative sign. In the case of the inverted pendulum, the acceleration is *away* from equilibrium instead of toward equilibrium like the harmonic oscillator.

If we look at eq. (4.6) we notice if we take the second time derivative of our function $\theta(t)$, we get that function back (multiplied by a constant). Recalling that $\frac{d}{dt}Ce^{\omega t} = \omega Ce^{\omega t}$, we might guess that $\theta(t)$ is an exponential.

Guess 1: $\theta(t) = Ce^{\omega t}$ **Guess 2:** $\theta(t) = Ce^{-\omega t}$

$\dot{\theta}(t) = \omega Ce^{\omega t}$ $\dot{\theta}(t) = -\omega Ce^{-\omega t}$

$\ddot{\theta}(t) = \omega^2 Ce^{\omega t}$ $\ddot{\theta}(t) = \omega^2 Ce^{-\omega t}$

$\ddot{\theta}(t) = \omega^2 \theta(t)$ $\ddot{\theta}(t) = \omega^2 \theta(t)$

Comparing to eq. (4.6), we see that both guesses work if the constant $\omega^2 = \frac{g}{L}$, so our solution to this second order differential equation is a linear combination of our two guesses:

$$\theta(t) = C_1 e^{\sqrt{g/L}\, t} + C_2 e^{-\sqrt{g/L}\, t} \tag{4.7}$$

where the constants C_1 and C_2 are determined from initial conditions. The following example illustrates the process to find C_1 and C_2.

Example. Model a dancer as a simple inverted pendulum with $L = 1.0\,\text{m}$. (The CM position is ~55 % of body height from the feet, so this dancer is ~182 cm or nearly 6 feet tall). If the dancer starts from rest ($\dot{\theta}_0 = 0$) from an initial topple angle θ_0, find $\theta(t)$ assuming small angles. Use the result to find the time to topple to a final angle of 15° from $\theta_0 = 0.1°, 1.0°$, and 5.0°. Compare the results to a shorter dancer with $L = 0.85\,\text{m}$ (~155 cm or 5′1″ tall).

Solution. When $t = 0$, eq. (4.7) becomes

$$\theta_0 = C_1 + C_2 \tag{4.8}$$

since $e^0 = 1$. Taking one time derivative of eq. (4.7) and setting $t = 0$ we have

$$\dot{\theta}_0 = \sqrt{\frac{g}{L}} C_1 - \sqrt{\frac{g}{L}} C_2 \tag{4.9}$$

Since for this example, $\dot{\theta}_0 = 0$, $\sqrt{\frac{g}{L}} C_1 - \sqrt{\frac{g}{L}} C_2 = 0$ and $C_1 = C_2$. And since $\theta_0 = C_1 + C_2$, this means that $C_1 = C_2 = \frac{\theta_0}{2}$, so we have

$$\theta(t) = \frac{\theta_0}{2}(e^{\sqrt{g/L}\, t} + e^{-\sqrt{g/L}\, t})$$

$$= \theta_0 \cosh(\sqrt{g/L}\, t)$$

where we have used the definition of hyperbolic cosine, $\cosh x = \frac{e^x + e^{-x}}{2}$. Solving for time we have

$$t = \sqrt{\frac{L}{g}} \cosh^{-1}\left(\frac{\theta}{\theta_0}\right)$$

We can now solve for the time to topple to our final angle, $\theta = 15°$ from different values of θ_0, given $L = 1\,\text{m}$ or $L = 0.85\,\text{m}$. The results are summarized in Tab. 4.1. Even for the very small initial angle of $0.1°$, both dancers would topple to $15°$ in less than two seconds. The $15°$ angle with the vertical, still seems like it may be relatively small; however to put that angle into perspective for balance, if a dancer is on flat feet and the heel to toe distance is 26 cm (US women's size 9.5 or men's size 9), the maximum topple angle (anterior–posterior direction) before it is no longer possible for the CM to be along a vertical line above the BoS is $15°$!

Table 4.1: Times for a taller ($L = 1.0\,\text{m}$) and shorter ($L = 0.85\,\text{m}$) dancer to topple from rest to an angle of $15°$ from three different initial topple angles (using the simple inverted pendulum model and small angle approximation).

Initial angle	$\theta_0 = 0.1°$	$\theta_0 = 1.0°$	$\theta_0 = 5.0°$
Time ($L = 1.0\,\text{m}$)	1.8 s	1.1 s	0.56 s
Time ($L = 0.85\,\text{m}$)	1.7 s	1.0 s	0.52 s

We could also find the time for the inverted pendulum to topple to $\theta = 90° = \frac{\pi}{2}$, or the time it would take for the dancer to fall until laying on the ground. We may no longer assume small angles in this case, so we solve the nonlinear differential equation, eq. (4.5) using a mathematical computational program (e. g., Mathematica's DSolve or NDSolve functions). The symbolic solution is not an elementary function (consisting of Jacobi amplitudes, the inverse of the elliptic integral of the first kind) so is not particularly illuminating. Instead, we numerically solve, providing the initial topple angle (θ_0) and initial topple velocity ($\dot{\theta}_0$). Results are graphed in Fig. 4.8 along with the solution using the small angle approximation. Note that there is very little difference between the exact and approximate solutions for these cases. In our hypothetical scenario, the dancer would topple all the way to the ground in less than three seconds from an initial position that is nearly vertical if the body is held perfectly rigid. Since dancers, and anyone who stands, remain upright for more than three seconds, balance maintenance is clearly a more dynamic process.

Example. Solve eqs. (4.8) and (4.9) to find C_1 and C_2 for *any* initial topple angle (θ_0) and topple velocity ($\dot{\theta}_0$).

Solution. Readers can do the algebra to show that $C_1 = \frac{1}{2}(\theta_0 + \sqrt{\frac{L}{g}}\dot{\theta}_0)$ and $C_2 = \frac{1}{2}(\theta_0 - \sqrt{\frac{L}{g}}\dot{\theta}_0)$. Substituting these results into eq. (4.7) and rearranging we have

$$\theta(t) = \frac{\theta_0}{2}(e^{\sqrt{g/L}\,t} + e^{-\sqrt{g/L}\,t}) + \frac{\dot{\theta}_0}{2}\sqrt{\frac{L}{g}}(e^{\sqrt{g/L}\,t} - e^{-\sqrt{g/L}\,t})$$

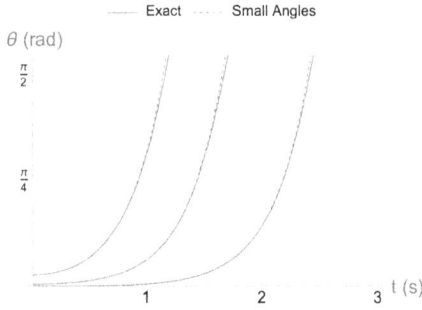

Figure 4.8: Topple angle as a function of time (L = 1.0 m) when toppling from rest from initial topple angles of 5°, 1°, and 0.1°, (from left to right). Exact solution (solid line) and solution assuming small angle approximation (dashed line) are shown.

and using the definitions of hyperbolic cosine and hyperbolic sine ($\sinh x = \frac{e^x - e^{-x}}{2}$) results in

$$\theta(t) = \theta_0 \cosh\left(\sqrt{\frac{g}{L}}\, t\right) + \dot{\theta}_0 \sqrt{\frac{L}{g}} \sinh\left(\sqrt{\frac{g}{L}}\, t\right) \tag{4.10}$$

4.4 Physical inverted pendulum

In our simple inverted pendulum model for balance, we have not yet included how the mass distribution of the body affects the moment of inertia of our system. To remedy this, we model the body as an inverted *physical* pendulum, which is an extended, rigid object. For example, if a dancer takes a very simple pose, e. g., standing on the toes (relevé parallel position) with the arms down, we could approximate the dancer's body as a single long cylinder (Fig. 4.9). The moment of inertia of the long cylinder (height $H = 2L$) around its end is $I = \frac{1}{3}MH^2 = \frac{1}{3}M(2L)^2 = \frac{4}{3}ML^2$, a larger inertia than if we modeled the dancer as a simple inverted pendulum with the mass at the CM ($I = ML^2$). So this physical pendulum model would lead to a smaller angular acceleration from balance, specifically $\ddot{\theta} \approx \frac{3}{4}\frac{g}{L}\theta$, compared to $\ddot{\theta} \approx \frac{g}{L}\theta$ for the simple pendulum. Since the torque due to gravity is always $MgL \sin\theta$ (where L is the distance of the CM from the point of support) and $I \propto ML^2$ no matter the shape of the physical pendulum, the equation of motion will always take the form

$$\frac{g}{cL}\theta = \ddot{\theta}$$
$$\frac{g}{L_{\text{eff}}}\theta = \ddot{\theta} \tag{4.11}$$

where c is a constant determined by the inertia of the physical pendulum, so we can define the effective pendulum length, $L_{\text{eff}} = cL$. In the end, the dynamics of toppling from balance are the same for the simple and physical inverted pendula, simply with a different effective pendulum length.

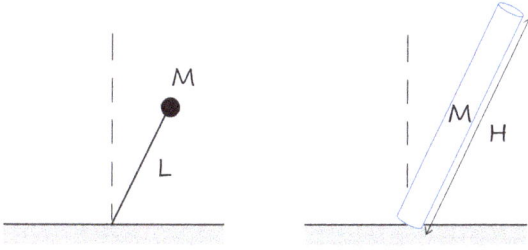

Figure 4.9: Inverted simple pendulum (left) and Inverted physical pendulum (right).

For more complex postures or for more fine tuned modeling, we could approximate the body as a compilation of various 3D shapes with known inertial properties (e. g., average inertial properties of body segments are given in the anthropometric data provided in [20]). A rudimentary model of a dancer standing with one leg extended to the side could include three long cylinders: one for the supporting leg, one for the extended leg, and one representing the head, arms, and trunk (HAT) segment. For more accuracy, and therefore complexity, we must acknowledge the fact that the inertia of a rigid body is not actually a scalar quantity, but a 3×3 tensor (which will be addressed in Chapter 5 on rotations).

Example. A dancer balances on the foot of one supporting leg (SL) with the gesture leg (GL) extended to the side (90° hip abduction). (a) Find the moment of inertia of the dancer around the topple axis approximating the head, arms, and trunk (HAT) segment as a cylinder with radius $r_{HAT} = 9$ cm, length $l_{HAT} = 73$ cm, and mass $m_{HAT} = 34.4$ kg, and each leg is a long cylinder with radius $r_{SL} = r_{GL} = 4$ cm, length $l_{SL} = l_{GL} = 100$ cm, and mass $m_{SL} = m_{GL} = 12.3$ kg (Fig. 4.10). (b) Compare your answer to part (a) to the inertia of the dancer standing on two feet (with the gesture leg parallel to the supporting leg).

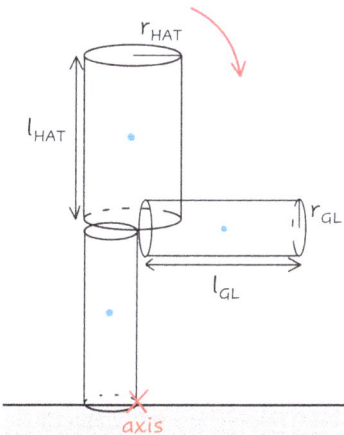

Figure 4.10: Simplified physical model of dancer with the gesture leg (GL) in second position. HAT = head–arms–trunk segment and SL = supporting leg.

Solution. (a) For this example, we start with the moment of inertia of a solid cylinder about an axis through its center and perpendicular to its longitudinal axis ($I = \frac{1}{4}mr^2 + \frac{1}{12}ml^2$), where r is its radius and l is its length). Then we use the *parallel axis theorem* to find its moment of inertia around an axis displaced a distance d from (and remains parallel to) the original axis ($I_\parallel = I + md^2$). The center of the HAT segment is displaced a distance $d_{HAT} = l_{SL} + \frac{1}{2}l_{HAT}$. And we use the Pythagorean theorem to find $d_{GL} = \sqrt{(\frac{l_{GL}}{2})^2 + l_{SL}^2}$ and $d_{SL} = \sqrt{(\frac{l_{SL}}{2})^2 + r_{SL}^2}$. We then have for the supporting leg

$$
\begin{aligned}
I_{SL} &= \frac{1}{4}m_{SL}r_{SL}^2 + \frac{1}{12}m_{SL}l_{SL}^2 + m_{SL}\left(r_{SL}^2 + \frac{l_{SL}^2}{4}\right) \\
&= \frac{5}{4}m_{SL}r_{SL}^2 + \frac{1}{3}m_{SL}l_{SL}^2 \\
&= 4.1\,\text{kg·m}^2
\end{aligned}
$$

for the gesture leg,

$$
\begin{aligned}
I_{GL} &= \frac{1}{4}m_{GL}r_{GL}^2 + \frac{1}{12}m_{GL}l_{GL}^2 + m_{GL}\left(\frac{5}{4}l_{GL}^2\right) \\
&= \frac{1}{4}m_{GL}r_{GL}^2 + \frac{4}{3}m_{GL}l_{GL}^2 \\
&= 15.4\,\text{kg·m}^2
\end{aligned}
$$

and for the HAT segment,

$$
\begin{aligned}
I_{HAT} &= \frac{1}{4}m_{HAT}r_{HAT}^2 + \frac{1}{12}m_{HAT}l_{HAT}^2 + m_{HAT}\left(l_{SL} + \frac{l_{HAT}}{2}\right)^2 \\
&= 65.7\,\text{kg·m}^2
\end{aligned}
$$

So, the total inertia of the body in this configuration is

$$
I = I_{HAT} + I_{GL} + I_{SL} = 86.2\,\text{kg·m}^2
$$

(b) If the GL were instead held parallel to the SL (zero degrees of hip ab/adduction), it would contribute the same amount to the total inertia as the SL found in part (a).

$$
\begin{aligned}
I &= I_{HAT} + I_{GL} + I_{SL} \\
&= (65.7 + 4.1 + 4.1)\,\text{kg·m}^2 = 73.9\,\text{kg·m}^2
\end{aligned}
$$

This is lower than the inertia found in part (a), which makes sense since more mass is further from the axis of rotation with the gesture leg lifted.

4.5 Quiet balance and dynamic balance

In the previous sections we've treated dancers as passive physical systems, allowing us to get a feel for how dancers would topple from an upright position if incapable of making any adjustments. Now we build some of the active nature of balance maintenance into our model, while keeping a good bit of its simplicity. Our simple inverted pendulum will have a simple modification: a static foot segment for the base of support and the axis of rotation at the ankle joint [108] instead of CP (Fig. 4.11). The force of gravity acts at the

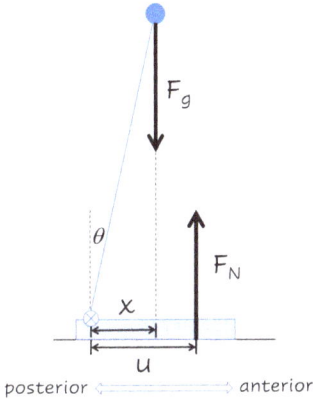

Figure 4.11: Inverted pendulum model with axis of rotation at the ankle joint. The CP can move in the x-direction so long as it remains within the BoS of the foot segment (in blue).

CM, located an effective pendulum length L_{eff} from the ankle joint. The GRF acts at the CP, which can now move anywhere within the BoS. With this model, dancers can use an "ankle strategy" to shift the CP in the anterior–posterior direction by activating the plantarflexors and dorsiflexors, producing a net muscle torque around the ankle joint [108]. When a dancer holds a pose, they often desire to hold it "quietly" with only ever so subtle adjustments and keeping the BoS fixed. The body remains relatively stationary during *quiet balance*, where the overall task goal is to keep the CM along a vertical line over the BoS.

With the axis of rotation at the ankle joint, two external torques act on the system, $\vec{\tau}_g$ and $\vec{\tau}_{\text{GRF}}$, so eq. (4.3) becomes $\vec{\tau}_g + \vec{\tau}_{\text{GRF}} = I\ddot{\theta}$. Let clockwise (CW) be the positive direction for θ and torque. Assuming quiet balance, θ and $\ddot{\theta}$ are relatively small, so $\overrightarrow{\text{GRF}} \approx F_N\hat{z} \approx Mg\hat{z}$. Since $\sin\theta = \frac{x}{L_{\text{eff}}}$ and for small angles $\theta \approx \frac{x}{L_{\text{eff}}}$, so $\ddot{\theta} \approx \frac{\ddot{x}}{L_{\text{eff}}}$. Applying Newton's second law for rotations (eq. (4.3)) we have

$$\vec{\tau}_g + \vec{\tau}_{\text{GRF}} = I\ddot{\theta}$$

$$Mgx - Mgu = ML_{\text{eff}}^2\left(\frac{\ddot{x}}{L_{\text{eff}}}\right)$$

$$\frac{g}{L_{\text{eff}}}(x - u) = \ddot{x} \tag{4.12}$$

This equation of motion for x is *very* similar to eq. (4.6) for θ and will therefore be easy to solve, but before doing so, let's first discuss balance maintenance with this model conceptually. Imagine a dancer standing on two feet in parallel position. Let's also assume for now that the projection of the CM remains anterior to the ankle joint. When $x > u$, eq. (4.12) shows that $\ddot{x}$ is positive, so the dancer's CM accelerates anteriorly. On the other hand, when $x < u$ as it is in Fig. 4.11, $\ddot{x}$ is negative, so the dancer experiences a torque to accelerate the CM posteriorly. The process of quiet balance maintenance is active because if the dancer's velocity is anterior, the ankle strategy can be used to move the CP more toward the toes (increasing u), causing a posterior acceleration [108]. Then as the velocity changes to moving in the posterior direction, the CP can be moved back toward the heels so the acceleration is anterior, and so on and so forth. This back and forth (and side to side) motion of the CM during quiet balance is termed *postural sway*. Although CM and CP motion are not necessarily the same, postural sway is often quantified in biomechanics research by use of tracking the CP position with a force plate or pressure mat.

Some studies have found differences in body sway parameters (e. g., CP displacement ranges, peak velocity, and trace area) between dancers and non-dancers during quiet balancing tasks. One study [46] found dancers to have lesser CP excursion (as quantified by the standard deviation of the CP trace) during quiet, single-leg standing than non-dancers who regularly exercise (Fig. 4.12). On the other hand, another study [21] found dancers had *greater* values of body sway parameters (CP displacement and ellipse area) than non-dancers during a single-leg stance. Traditionally, increased sway is interpreted as inferior balancing ability compared to lower sway values, but caution should be made when using this interpretation for dancers. Larger amplitudes of CP movement within the BoS may be exploratory as a dancer feels for balance [21] or may be indicative of dancers having greater strength to withstand larger displacements from equilibrium while remaining upright. Yet another study found that dancers' CP motion was noisier than other athletes, but that the dancers' postural motion occurred around a more constant mean position, which the authors suggest may indicate "behavioral flexibility" in dancers' balance maintenance [89].

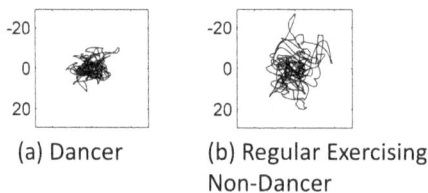

(a) Dancer

(b) Regular Exercising Non-Dancer

Figure 4.12: 2-D center of pressure trace during a 20-second, quiet single-leg stance for (a) a dancer and (b) a regularly exercising student. Units are in mm. Reprinted with permission from [46].

Dance is much more than holding quietly balanced poses, however, and dancers must remain balanced during more dynamic situations. For example in a piqué arabesque, a dancer launches the body over the supporting leg which is held straight, giving themselves forward momentum that then must decrease as they enter the final, balanced arabesque position (Fig. 4.13). There are many examples in dance where a dancer must have an appropriate amount of initial momentum to launch the CM over the BoS as they enter a pose or a move, whether the BoS is the ball of one foot, like the piqué arabesque or the hands, as in an inversion like a handstand. Too little momentum and the CM will not reach a position over the BoS. Too much momentum and the dancer's CM will overshoot the BoS.

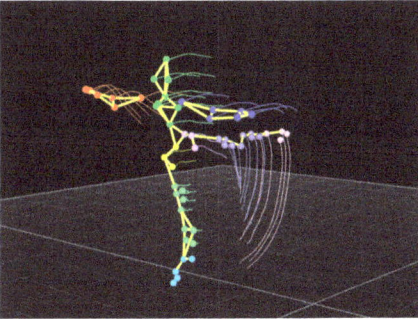

Figure 4.13: A dancer does a piqué arabesque. The final balanced arabesque position is shown, and the colored lines are the marker trajectories in the preceding frames as the dancer moves into arabesque.

Pai and Patton [74] recognized the importance of both the CM position and velocity when it comes to *dynamic balance* maintenance. Even if the CM is over the BoS, balance may not be possible to maintain if the CM velocity is too great. Following the inverted pendulum model in Fig. 4.11 they predicted the combinations of initial position and initial velocity that would allow the CM to come to rest at a final position vertically above the BoS. Their model was subject to reasonable physical constraints (F_N must be vertically upward, $F_f < F_{s,\max}$ so the foot did not slip, and the CP remains in the BoS) and physiological constraints (the ankle torque stayed within the range possible based on maximum muscle strength). Fig. 4.14 displays the possible x_0 and $\dot{x}_0$ combinations (shaded region) in the anterior–posterior direction for dynamic balance maintenance.

Returning to eq. (4.12) and solving for x will additionally highlight the importance of both x_0 and $\dot{x}_0$ for balance maintenance. We notice eq. (4.12) is nearly identical to our previous equation of motion for θ (eq. (4.5)) for the simple inverted pendulum, with the only difference being the extra constant term u. In fact, if we define a new variable $w(t) = x(t) - u$, we can rewrite eq. (4.12) as $\frac{g}{L_{\text{eff}}}(w) = \ddot{w}$, and the solution for w is the same as the solution for θ (eq. (4.10)). So,

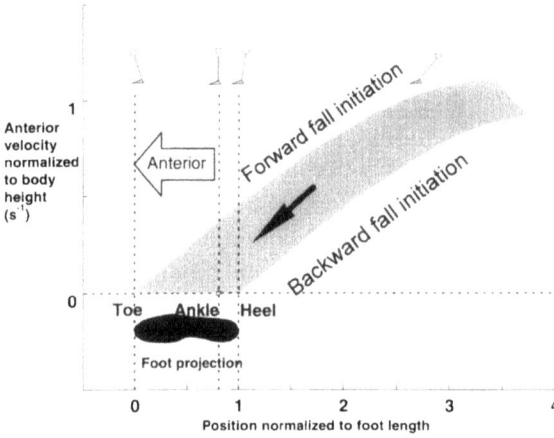

Figure 4.14: Allowable anterior initial velocity and initial position combinations for the CM to come to rest over the BoS (shaded region). Forward falls occur if the x_0 and $\dot{x}_0$ fall in the region above the upper boundary and backward falls occur if they are below the lower boundary. Position and velocity are normalized such that x_0 is given as a number of foot lengths and $\dot{x}_0$ is fraction of body height per second. Reprinted with permission from [74].

$$w = w_0 \cosh\left(\sqrt{\frac{g}{L}}\,t\right) + \dot{w}_0 \sqrt{\frac{L}{g}} \sinh\left(\sqrt{\frac{g}{L}}\,t\right)$$

$$x(t) - u = (x_0 - u_0) \cosh\left(\sqrt{\frac{g}{L}}\,t\right) + \dot{x}_0 \sqrt{\frac{L}{g}} \sinh\left(\sqrt{\frac{g}{L}}\,t\right) \tag{4.13}$$

Hof, Gazendam, and Sinke [34] took the result in eq. (4.13) and defined a new condition for dynamic stability, recognizing that for an acceleration back toward equilibrium (or zero acceleration) $x \leq u$ (Fig. 4.11), so from eq. (4.13),

$$0 \geq (x_0 - u_0) \cosh\left(\sqrt{\frac{g}{L}}\,t\right) + \dot{x}_0 \sqrt{\frac{L}{g}} \sinh\left(\sqrt{\frac{g}{L}}\,t\right)$$

$$u - x_0 \geq \dot{x}_0 \sqrt{\frac{L}{g}} \tanh\left(\sqrt{\frac{g}{L}}\,t\right)$$

And since $\tanh x$ takes a range of values from -1 to 1, $u - x_0 \geq \dot{x}_0 \sqrt{\frac{L}{g}}$ or

$$u \geq \dot{x}_0 \sqrt{\frac{L}{g}} + x_0 \quad \text{(Condition for dynamic stability)} \tag{4.14}$$

Following this reasoning, it is not just the initial position, x_0 that must be within the BoS for balance, but the quantity $\dot{x}_0 \sqrt{\frac{L}{g}} + x_0$ [34]. The same argument follows for CM motion in the medial-lateral direction (y), so we can conclude more generally that $\vec{u} = \dot{\vec{x}}_0 \sqrt{\frac{L}{g}} + \vec{x}_0$ for dynamic balance, where $\vec{u} = u_x \hat{x} + u_y \hat{y}$ and $\vec{x}_0 = x_0 \hat{x} + y_0 \hat{y}$.

Example. Find the range of initial anterior velocities that allow a dancer to reach the final piqué arabesque position on demi-pointe with the CM at rest over the BoS. Assume that the length of the BoS is 7 cm, the initial CM position is 20 cm behind (posterior to) the center of the BoS, and $L_{eff} = 1.0$ m.

Solution. Let the center of the BoS be at the position $x = 0$. The maximum and minimum positions of the CP are therefore $u_{max} = 3.5$ cm and $u_{min} = -3.5$ cm and $x_0 = -20$ cm. Using eq. (4.14) we solve for the maximum and minimum possible initial CM velocity values:

$$u_{max} = \dot{x}_{0,max} \sqrt{\frac{L}{g}} + x_0 \qquad\qquad u_{min} = \dot{x}_{0,min} \sqrt{\frac{L}{g}} + x_0$$

$$0.035\,\text{m} = \dot{x}_{0,max} \sqrt{\frac{1.0\,\text{m}}{9.8\,\text{m/s}^2}} - 0.20\,\text{m} \qquad -0.035\,\text{m} = \dot{x}_{0,min} \sqrt{\frac{1.0\,\text{m}}{9.8\,\text{m/s}^2}} - 0.20\,\text{m}$$

$$\dot{x}_{0,max} = 0.74\,\text{m/s} \qquad\qquad \dot{x}_{0,min} = 0.52\,\text{m/s}$$

For a dynamically balanced, piqué arabesque, this dancer needs an initial CM velocity between 52 cm/s $\leq \dot{x}_0 \leq$ 74 cm/s. This relatively large range of velocities makes it such that the dancer need not be too precise with the initial velocity; however, a complicating factor with the piqué arabesque is that the gesture leg is lifted anteriorly using hip extension as the dancer moves into the final balanced position. This changes the CM position relative to the dancer's body, so it does take practice to get a feel for the appropriate amount of initial velocity. Dancers can also alter the supporting foot placement as they prepare for the piqué arabesque, making x_0 larger or smaller depending on how they step. It is also true in other dynamic situations, such as kicking into a handstand, that both the appropriate BoS placement (setting x_0) and launch velocity ($\dot{x}_0$) are important to successfully allow for a final balanced pose.

4.6 Double and triple inverted pendulum and kinematic strategies for balance maintenance

The single inverted pendulum model is useful when assessing the relative motion between the CM and CP in the overall task goal for balance maintenance. However, how we physically *control* the positions of the CM and CP requires additional complexity in our model. The single inverted pendulum assumes our only defense against the torque due to gravity is via joint actions at the ankle, while the rest of the body is held rigidly. During more challenging postures (like standing on one foot) or when large perturbations from balance occur, other joints are often involved. If a dancer is falling forward, a "hip strategy" may be utilized by leaning forward with hip flexion (Fig. 4.15). It may seem counterintuitive to lean in the direction of topple; however, this motion leads to an external force on the dancer in a direction back toward equilibrium. We can understand how that is with the following reasoning. Using the hip strategy, the hip flexors produce a torque causing the legs to rotate around the hip (CCW in Fig. 4.15), and this causes the feet to exert a force on the ground in the anterior direction. By Newton's third law, the ground exerts a force on the dancer (horizontal GRF) in the posterior direction, back toward equilibrium [57]. Fighting against an imbalance by leaning the body away from the direction of topple can actually make matters worse by directing the GRF away from equilibrium.

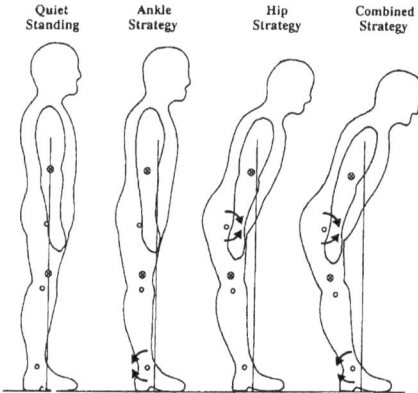

Figure 4.15: Examples of strategies for upright balance maintenance. Reprinted with permission from [108].

In general, other joint actions or combinations of joint actions may be used during balance maintenance, like the combined hip–ankle strategy shown in Fig. 4.15. Recall that the dancer in the beginning of chapter example "did a graceful arm movement" that was likely used as a combined joint strategy. Studies have shown that even for quiet balance without large disturbances from equilibrium, multiple joint actions are utilized simultaneously during balance maintenance. The number of joint actions involved has been shown to increase as the postural challenge increases (e. g., standing on one leg vs. two) as well as in dynamic balance situations [45]). Instead of a single inverted pendulum with the only joint actions at the ankle, in this case kinematics of balance maintenance would be better modeled with a multi-segment inverted pendulum to include additional joint actions. A double inverted pendulum with one segment for the lower extremities and one segment for the head–arms–trunk connected via an axis at the hip would better capture the combine ankle-hip strategy. A triple inverted pendulum could include segments for the lower legs, the upper legs, and the head–arms–trunk segment with joints at the ankle, knee, and hip (Fig. 4.16).

One study using the model in Fig. 4.16 found that during quiet standing on two feet, there were no significant differences in the sway amplitudes at the ankle, knee, and hip

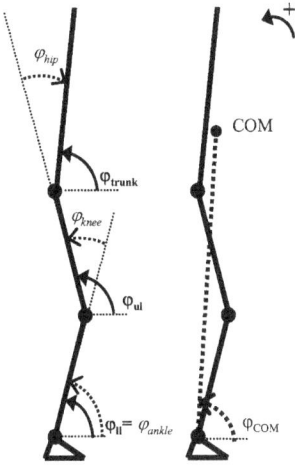

Figure 4.16: Triple inverted pendulum for kinematic coordination patterns between the ankle, knee and hip joints. Reprinted with permission from [79].

joints (measured by the standard deviation of the joint angles, ϕ_{ankle}, ϕ_{knee} and ϕ_{hip}) [79]. This demonstrates that even for quiet standing, the hips and knees are not held rigidly. If all motion only occurred around the ankle joint with the knees and hips held locked with $\phi_{\text{knee}} = \phi_{\text{hip}} = 0$, then the angles that the segments make with the horizontal (ϕ_{ll}, ϕ_{ul}, and ϕ_{trunk}) would remain the same for all times. If this were the case, the plots in Fig. 4.17 would be perfectly straight lines, which they are not. The researchers found that the amount of deviation from a straight line varied between individuals in the study (with correlation coefficients ranging from 0.3 to 0.98). Interestingly, they found that the CM angle and ankle angle still had overall displacements of similar magnitude, and that there was a countermovement between the HAT and lower leg segment, meaning these segments swayed in a coordinated manner but out of phase.

Simply studying joint angular displacements does not provide full information about control of balance. While the amplitudes of joint angles at ankle, hip, and knee may be similar during quiet stance, the net joint torques due to muscle activation around the ankle joint are often larger due to the joint's proximity to the support surface in an upright posture. Some joint actions, especially those further from the point of contact with the support surface, may be a consequence of inertial effects extending outward from joints closer to the pivot point. Understanding balance control requires investigating which muscles are active, when and how neurologically they are activated, which we will discuss in the following section.

4.7 Sensing balance and neuromuscular control

When maintaining any balanced pose, two distinct task goals are present, each with different muscular involvement and neural circuitry. The first task is to hold the "shape" of the pose (otherwise known as the "posture"). For example, a handstand and an

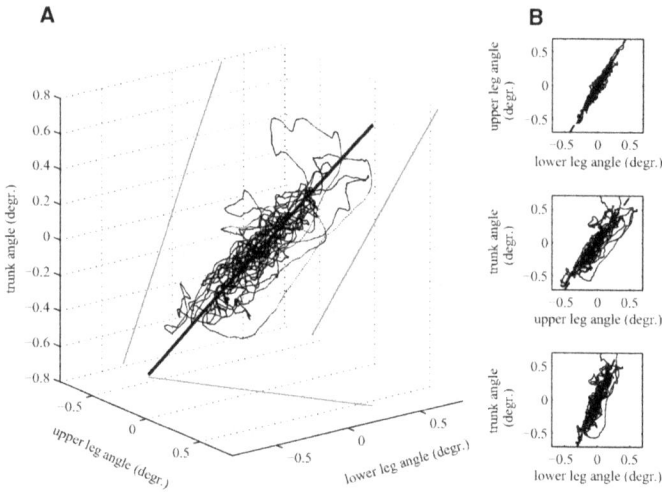

Figure 4.17: A 3D plot of the segment angles (A) and their projections on the 2D planes (B). Reprinted with permission from [79].

arabesque are two poses that each require different sustained muscle activity to simply hold the positions of the body's segments in fixed positions relative to one another. A dancer can feel the muscles in the glutes and back continuously active to maintain the gesture leg extended posteriorly in an arabesque. Even simply standing on two feet requires continual low-level muscle activity to hold the posture with the knees extended and the trunk vertical.

The second task, which is our primary focus, is to manage the torque due to gravity via the relative positioning of the CM and CP. The muscles elicited and neural control to compensate for deviations from equilibrium are generally different from those to maintain a fixed posture. During balance maintenance, sensory information is gathered relevant to the balancing conditions, that information is sent to the CNS, and motor outputs are transmitted to activate the muscles, eliciting kinematic strategies to move the CP and/or CM in the overall task goal.

Let's again consider our dancer from the beginning of the chapter in regards to timing of the sequence events outlined above. When the dancer experiences a perturbation from equilibrium, there is a brief *sensory delay* between the actual motion (when the dancer is shoved) and the threshold at which the sensory receptors detect the disturbance (t_1 in Fig. 4.18 [11]). Then it takes additional time for that sensory information to be transmitted to the CNS, for the CNS to process it, and for motor signals to be sent back to the muscles. This time between sensory detection and subsequent muscle activation is called the *neurological delay* (from t_1 to t_2 in Fig. 4.18). Then there is a brief *electromechanical delay* between when the muscles are activated and when they begin generating force. Those muscle forces produce the "strategy" of joint torque(s) that move the CP and/or CM. The sensory, neurological, and electromechanical delays all depend

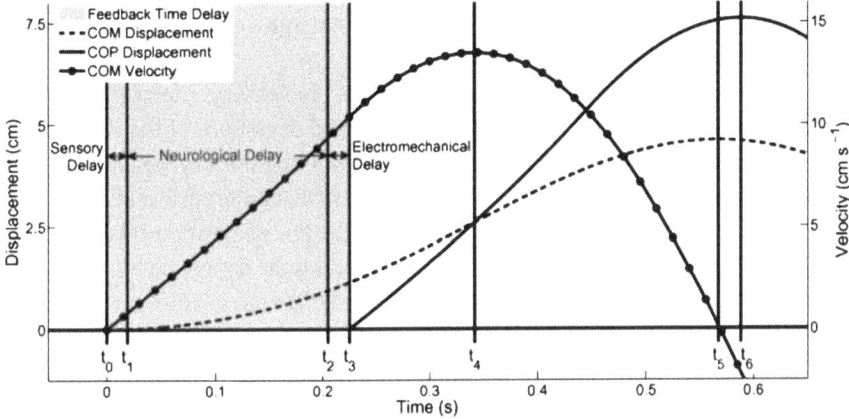

Figure 4.18: An example of the timing between when a perturbation from balance occurs (t_0), the sensory threshold is reached (t_1), the muscles are activated (t_2), and muscle force is produced (t_3). In this example, the muscle force initiates a displacement of the CP and at t_4 it crosses the CM to produce a restoring force accelerating the CM toward equilibrium. Reprinted with permission from [11].

on various factors, such as the individual, the specific balancing task, and conditions. For example, the neurological delay is generally longer when higher order CNS processing is involved versus for reflexes. In general, the overall delay between an unexpected perturbation from balance and muscle force generation is ~0.1 to 0.2 seconds, however, the timing of the delays differ between quiet balance and perturbed balance or if a perturbation is expected, muscles may generate anticipatory responses.

The neural pathways and areas of the CNS involved to maintain postural equilibrium depend on the posture and the level of challenge to balance. In general, contemporary research has shown that maintaining a balanced pose is more of a motor skill than a set of reflexes [39], and greater postural challenges involve higher levels of cortical involvement. For a simpler task like standing on two feet, motor control of balance seems to lie somewhere between a more rudimentary stretch reflex and higher-order control involving the motor cortex of the brain. In this situation, postural equilibrium is maintained via *automatic postural responses* (APR), which, like reflexes, are involuntary and short latency (shorter neurological delay), but the neural circuitry is more complex than that of reflexes. The multiple muscles involved in an APR are called a postural set. Additional research has found that a relatively small set of neural commands form the basic building blocks that produce overall muscle coordination for APRs in postural tasks [98]. These *muscle synergies* are patterns of muscle activation elicited by the nervous system to achieve a specific task goal like balance maintenance [98]. Research suggests that some of these muscle synergies may be innate, but differences between individuals in muscle synergy patterns and the number of synergies each individual possesses suggest that muscle synergies may be adaptable and/or new synergies may form with training [99]. There is not merely one possible combination of muscle synergies for balance maintenance; instead, multiple combinations of synergies could produce the same overall kinematics in moving the CM and/or CP.

When holding a pose, a dancer continuously takes in sensory information, evaluating both the relative positioning of body segments and deviations of the whole body from equilibrium. The three areas most involved in providing sensory feedback for balance maintenance are the visual system, vestibular system, and proprioceptive system. Our vision provides one mechanism for monitoring the body's position relative to another object, so a dancer keeping a gaze on a fixed "spot" (as the dancer mentioned at the beginning of the chapter) can be a useful technique to assist with balance maintenance. Both dancers and non-dancers exhibit better postural stability in eyes open versus eyes closed balancing tasks [76], but some research indicates that dancers rely less on visual feedback than non-dancers [93]. Dancers might use a mirror present in a dance studio to find their spot and other reference points when dancing, however there are drawbacks to heavy reliance on a mirror including that the landmarks a dancer may use in the dance studio will be unavailable in a performance space without a mirror.

The *vestibular system* of the inner ear consists of 3 orthogonal, fluid-filled canals approximately aligned with the body's sagittal, frontal, and transverse planes. Tiny sensory hair cells line the canals and sense the motion of the fluid when the head moves. The vestibular system provides sensitive feedback about the position and orientation of the head as well as its motion. Spinning can compromise vestibular feedback and is why we may experience a dizzy feeling after rotational movement. Conflicting information between the vestibular system, vision, and proprioception can affect our ability to maintain balance and is also the reason why some people experience motion sickness while traveling.

Proprioceptors, as we learned in Chapter 2, are sensory receptors within joints, muscles, and tendons that sense mechanical properties such as position, velocity, or force. For example, if a dancer begins to lean anteriorly, muscle spindles in the posterior leg muscles (e. g., calf muscles) can send information to the CNS about this increase in muscle length or stretch. Cutaneous receptors located in the skin act as pressure sensors, so dancers can feel the CP shift as the pressure distribution over the BoS changes.

All of the sensory information, when sent to and combined in the CNS, provides an overall picture of the dancer's body position, orientation, and motion and this representation is evaluated in reference to the overall task goal (Fig. 4.19). How the CNS weights the various forms of sensory feedback depends on multiple factors, such as the specific task and conditions (e. g., magnitude of displacement from equilibrium), availability of sensory information (e. g., if the eyes are open or closed), and an individual's training [7]. While visual and vestibular information may be useful to dancers for evaluating balance, studies have provided evidence that dance training increases reliance on proprioceptive feedback [92, 93, 8]. Why might this happen? Dance pedagogy often encourages dancers to evaluate how movements *feel* when performing them as opposed to simply evaluating the external quality of the movement itself. This overall movement sense, or kinesthia, can become highly developed in dancers. Research has also shown

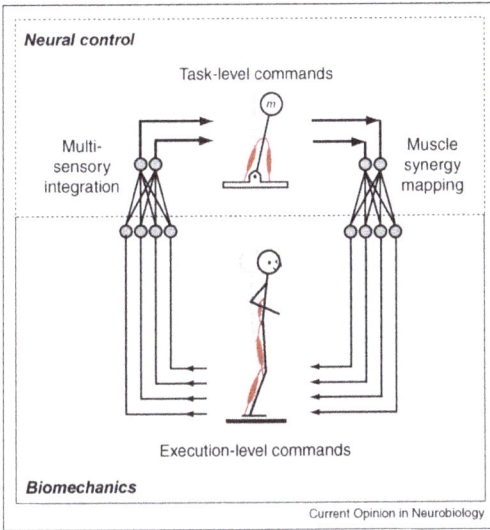

Figure 4.19: Neuromechanics of balance maintenance. Sensory feedback is weighted and integrated in the CNS, where the dancer's posture is evaluated in reference to the overall task goal, then the goal is mapped onto a set of muscle synergies to produce the motor output. Reprinted with permission from [98].

that dancers can be specifically trained to develop proprioceptive strategies for balance maintenance with eyes closed dance training [38]. Increased reliance on proprioception over vision can give dancers an edge when performing in a darkened theater and in new environments or unpredictable conditions.

Factors such as injuries or joint hypermobility may affect a dancer's proprioceptive abilities. Soft tissue injuries may compromise the sensory receptors at the injury site, and it has been shown that ankle sprains lead to decreased proprioception in dancers [8]. Because proprioception is so important for dancers' balance, part of their rehabilitation from injury should include exercises to promote proprioception like balancing on wobble boards or balance balls. Even without a prior injury, dancers may have deficits in any one of their sensory feedback systems that could increase their risk of injury. Screening dancers for these deficits by using standard balance tests or developing dance style-specific balance tests would be a beneficial part of overall dance wellness protocols. Increased hypermobility in non-dancers has been shown to have a negative effect on proprioception, so it may be expected that dancers would experience similar results. While one study found greater ankle joint laxity negatively affected dancers' dynamic postural stability [71], results of other studies suggest balance ability and proprioception in dancers to be unaffected by hypermobility [67] or that hypermobility was *positively* correlated with balance test scores in dancers [4]. The effects of different types of stretching on balance performance has also been studied. A combination of static and dynamic stretching as part of a warm-up was found to be more beneficial in terms of dancers' balance performance than static stretching alone [72].

4.8 Upright versus inverted balance maintenance

In our analyzes of balance maintenance, so far we have focused mainly on upright postures with the base of support between the feet (or one foot) and the floor. However, the same basic principles for postural equilibrium apply to inverted postures, which use the upper extremities as supporting limbs. Inversions are prevalent in gymnastics in addition to dance forms such as breaking and acrobatics, for example. The relative positioning of the CM and CP for balance maintenance in our simple inverted pendulum model still apply to inversions, as well as the condition for dynamic stability in eq. (4.14).

Quiet balance on two hands has some parallels with quiet balance on two feet, but some important differences make inverted balance maintenance substantially more challenging. Although the CM may be further from the ground in an inverted stance, which would enhance stability, the handstand is biomechanically less stable than standing on two feet for multiple reasons. Standing on the hands often has a smaller BoS than standing on two feet. However, standing on one foot or on the balls of two feet (e. g., in ballet fifth position relevé) can actually have a smaller BoS than a standard handstand. So why is a handstand a generally more challenging posture than fifth position relevé? The primary reasons are neurological and muscular. For one thing, the head is closer to the pivot point in the handstand, so the amplitude of the head's motion will be less than that in an upright posture. This means that the vestibular system and visual systems (with sensors only in the head) will not be as effective in providing sensory feedback. Another neurological consideration is that the muscle synergies and APRs that exist for upright postures are not applicable to inverted postures and it would take time and extensive practice to develop postural sets relevant for an inverted posture. Additionally, strength of the muscles in the upper limbs is typically not as great as those in the lower extremities, so it is more challenging to generate the muscle torques necessary (e. g., wrist flexion) to maintain an inverted posture.

Despite these differences, research has demonstrated some similarities between upright and inverted balance maintenance. Like upright balance, kinematic coordination between multiple joints are present in the handstand. Research has found that wrist, shoulder, and hip joint torques all contribute to CM movement during the handstand, with the wrist generally playing the most important role (similar to the ankle joint in upright balance) [44, 49]. Elite gymnasts have been demonstrated to rely on activation of the wrist flexors more than less experienced gymnasts (who emphasize more activation at the shoulders and hips along with the wrists) for postural control [49]. One study asked gymnasts (all classified as experts specifically in the handstand) to perform a dynamic balancing task: tracking a target with their ankles during a handstand that oscillated in the anteroposterior direction [28]. The researchers found differences in overall performance and the joint strategies used depending on whether or not the participants were also classified as general gymnastics experts ("high expert") or if they were only specifically expert in the handstand ("low expert") (Fig. 4.20). The "high experts" performed better at the dynamic balance task and utilized a strategy dominated by the wrists and

High Expert Low Expert

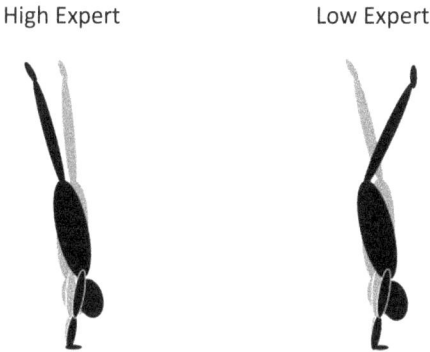

Figure 4.20: Illustration of different coordinations used by gymnasts when tasked to track a target with their ankles (oscillating in the anteroposterior direction at 0.4 Hz). "High expert" = high expert level in the handstand specifically and gymnastics in general vs "Low expert" = high expert level in the handstand but intermediate level in gymnastics in general. Reprinted with permission from [28].

synchronized, small-amplitude motion of the shoulders and hips, compared to the "low experts," whose coordination was led by the hips, which oscillated out of phase with the shoulders. The researchers surmised that the general expertise of the high experts enabled them to be more adaptable to slight modifications to the specific task.

4.9 Balance and dance training

Many varied quiet and dynamic balance situations exist in dance that utilize vastly different postures and base of support configurations. The various postures may rely on different combinations of muscles for balance maintenance, depending on the specific shape of the body configuration and whether postural sway is primarily in the anterio-posterior, mediolateral, or a combined direction. Some dance styles, like contemporary, place greater emphasis on complex and varied poses than other styles, but it may not be necessary for these dancers to specifically train every possible posture for balance. The study on inverted balance maintenance comparing "high" and "low" experts provided evidence of transferability of similar skill-sets in the general high experts, allowing them to be more adaptable to modifications than skill-specific training. This also supports the notion that dance training may improve balance maintenance in general.

In the previous section, we discussed how dance training can lead to re-weighting of sensory information, and an increase in reliance on proprioception, used during balance maintenance. In response to perturbation from balance, one study found ballet dancers have significantly faster and more consistent muscle activation compared to non-dancers [92]. It also appears that resistance to fatigue and vestibular stimulation (e. g., spinning in pirouettes) on balance test performance increases with dance training [6, 37]. Dance training may also change the kinematics and/or motor outputs one uses to maintain balance. One study in particular compared dancers and non-dancers in dynamic balance tasks (normal walking over ground and walking on a balance beam, which neither the dancers or non-dancers were specifically trained to do). The researchers found that although the leg and trunk kinematics were similar between groups, the dancers used a greater number of muscle synergies, were more consistent in their muscle activation, and used less coactivation (and were therefore more efficient) than the non-dancers during walking on the beam [87]. They also found greater similarity in muscle synergies used in overground and beam walking for the dancers than non-dancers. The results led the authors to conclude that the years of dance training (ballet in this case) changed muscle patterns and neural control of even the everyday task of overground walking, and their expanded motor pool could explain why dancers are often well-equipped to adapt to challenging balancing tasks like beam walking. The dancer at the beginning of the chapter is likely a "high expert" in upright balance whose many years of dance training have changed her muscle synergies and neural control. Her rapid responses and increased motor repertoire allow her to adapt seemingly effortlessly to many challenging balance scenarios.

It is not necessary for one to study dance for many years or at the professional level to experience benefits in regards to balance. For example, dance has been studied as an intervention or therapy for patients with neurological disorders, such as cerebral palsy, Parkinson's disease, and multiple sclerosis. Research has shown that dance programs of various styles (e. g., different types of ballroom, ballet, aerobic, and social or cultural dance) for ranges weeks, as opposed to years, have led to improvements in patients' balance outcomes (as well as other biomechanical and cognitive benefits) [53, 69, 88]. Dance for Health as a field has been growing with promising results, but its discussion more broadly is beyond the scope of this book.

5 Rotations

A dancer practices triple pirouettes over and over without improvement. He places his feet in fourth position to prepare for the turn, then pushes off and onto a single supporting foot as his body spins around a vertical axis. But no matter how many times he tries, he is unable to make it around for the three full rotations without hopping on his supporting foot or an off-balanced, clumsy landing of the turn. The frustrated dancer is unsure of what to do except to keep trying the same approach repeatedly. The dancer wonders, "Why can I do a double pirouette with ease, but am having so much trouble just squeezing in one more revolution? Should I change my push-off for the turn? Should I focus more on my spot, keeping my eyes on a fixed point on the wall after each revolution? How can I land the turn more gracefully?" Luckily for this dancer, there are physical and biomechanical concepts that can help answer these questions and a growing body of research on the biomechanics of whole-body dance rotations. While much of the research has focused on turns in ballet, many of the ideas can be generalized to other dance forms (and rotations outside of dance more broadly).

Much of this chapter focuses on rigid body rotations. A *rigid body* is an object with a fixed shape; in other words, the distance between any two particles that make up the object remains fixed. Clearly, this definition does not always apply to the whole dancing body. There are, however, some situations when there is little relative motion between a dancer's body parts and a rigid body approximation for the dancer's whole body is justified. In many other scenarios, we can approximate certain parts of a dancer's body as rigid, like the leg segment as it rotates about the hip joint in a kick.

The idea of rigid bodies is important when expanding our analyses of dance movement to include motion of parts of the body relative to the center of mass (CM). Dancers are anything but point particles, and in dance, the movement of the rest of the body relative to the CM is just as (if not more) important than CM movement. Whether a dancer is executing a whole-body rotation like a pirouette or the arm rotates around the shoulder joint in a port de bras, rotations are abundant in dance and human movement more generally. When it comes down to it, all human movement is accomplished by rotating body segments around their respective joints.

5.1 Rotational kinematics

In the previous chapters, we used linear kinematics and dynamics to describe how and why the CM experiences translational movement. *Rotational* kinematics and dynamics allow us to describe how and why rigid bodies rotate. Let's consider a dancer spinning in a pirouette after reaching the fixed "pirouette position," whatever body position that may be. For now, we will approximate the whole dancer as a rigid body with no relative motion between body segments. Let's imagine the dancer rotates around a vertical axis and that the CM remains aligned over the center of pressure. In this case, there is

https://doi.org/10.1515/9783110642292-005

no translational motion of the CM, but there is rotational motion of the body around an axis through the CM. For the dancer performing the pirouette, the global coordinate system (*xyz*) remains fixed in space, but the local (body) coordinate system of the dancer (*X'Y'Z'*) is fixed relative to the dancer. For the body coordinate system, we will use our familiar anatomical axes: the mediolateral (*X'*), anteroposterior (*Y'*), and longitudinal (*Z'*) axes. As the dancer rotates around the vertical (*z* and *Z'*), the body coordinate system rotates with the dancer. (In later sections, we will consider more general rigid-body rotations, in which the *xy*- and *X'Y'*-planes do not remain parallel during the turn.)

To measure the pirouette's rotational kinematic quantities, let's imagine a top-down view of the rotating dancer (Fig. 5.1). We will define the *rotational position* ($\vec{\theta}$) as the angle between *X'* and *x*, and let the direction of $\vec{\theta}$ be positive when measured counterclockwise. Standard units of measurement for θ are radians (rad). *Rotational velocity* ($\vec{\omega}$) is the time rate of change of rotational position

$$\vec{\omega} = \frac{d\vec{\theta}}{dt} = \dot{\vec{\theta}} \tag{5.1}$$

and describes how rapidly the dancer rotates and in what direction (positive rotational velocity means rotating counterclockwise and negative means rotating clockwise) with units of rad/s. *Rotational acceleration* ($\vec{\alpha}$) is the time rate of change of rotational velocity

$$\vec{\alpha} = \frac{d\vec{\omega}}{dt} = \ddot{\vec{\theta}} \tag{5.2}$$

and describes how rapidly the rotational velocity changes and in what direction (CW or CCW) with units of rad/s^2.

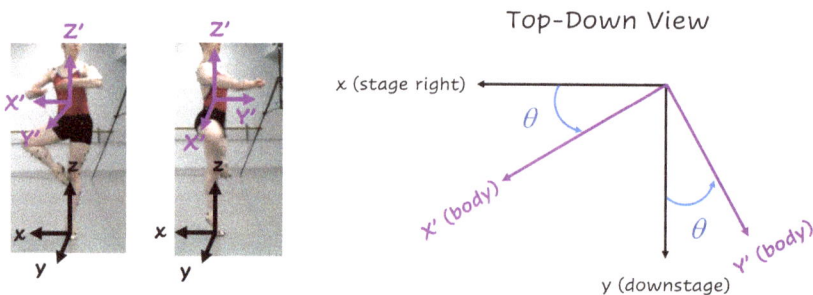

Figure 5.1: Dancer's body coordinate system (*X'Y'Z'*) and the fixed global coordinate system (*xyz*) as a dancer rotates around the vertical in an en dedans pirouette (spinning counterclockwise around the vertical). Top-down view shows the *x, y* and *X', Y'* axes at a snapshot in time between the two photos to the left. (*z* and *Z'* are out of the page in the top down view.) θ is defined as the angle that *X'* makes with *x*, and is the same angle that *Y'* makes with *y*.

If rotational acceleration is constant, the kinematics equations with which we are familiar can be applied to rotational motion by substituting the linear quantities with their respective rotational quantities:

$$\vec{x} = \vec{x}_0 + \vec{v}_0 t + \frac{1}{2}\vec{a}t^2 \longrightarrow \vec{\theta} = \vec{\theta}_0 + \vec{\omega}_0 t + \frac{1}{2}\vec{\alpha}t^2$$

$$\vec{v} = \vec{v}_0 + \vec{a}t \longrightarrow \vec{\omega} = \vec{\omega}_0 + \vec{\alpha}t$$

$$\vec{v}^2 = \vec{v}_0^2 + 2\vec{a}\Delta\vec{x} \longrightarrow \vec{\omega}^2 = \vec{\omega}_0^2 + 2\vec{\alpha}\Delta\vec{\theta}$$

Example. As a dancer spins on one foot, friction between her shoe and the ground causes her rotational velocity to decrease from 3 rev/s to 2 rev/s after her body completes one full revolution. Determine the dancer's rotational acceleration (assumed constant) from this information.

Solution. We could begin by converting the rotational kinematic quantities to standard units of rad and rad/s to obtain rotational acceleration of rad/s^2; however, if we keep our units consistent when applying a kinematics equation, there is no need to change units. We will instead obtain a rotational acceleration in units of rev/s^2. Given what we know, we can solve for a with a straightforward application of the third listed above kinematics equation.

$$\omega^2 = \omega_0^2 + 2\alpha\Delta\theta$$

$$(2\,\mathrm{rev/s})^2 = (3\,\mathrm{rev/s})^2 + 2\alpha(1\,\mathrm{rev})$$

$$\alpha = -\frac{5}{2}\,\mathrm{rev/s}^2$$

The negative sign indicates that the rotational acceleration is in the direction opposite to the dancer's rotational velocity, decreasing her rotation rate.

With the whole body in a fixed position in a rotation (e. g., arabesque turn, with the gesture leg extended posteriorly), all points on the dancer's body travel the same rotational displacement ($\Delta\theta$) in a given amount of time, so the rotational velocity is the same for every point on the body. However, points on the dancer's body further from the axis of rotation will travel a greater overall distance. The circumference of the circular path traced out by the dancer's foot in arabesque is greater than that traced out by the knee during one revolution of an arabesque turn. This also means that the tangential velocity of the foot is greater than the tangential velocity of the knee in this example. The relationships between linear and rotational kinematic quantities can be found beginning with arc length (s), and noting that r is the distance of the point of interest from the axis of rotation (assumed constant).

$$s = r\theta$$

$$v_t = \frac{ds}{dt} = r\frac{d\theta}{dt} = r\omega \tag{5.3}$$

$$a_t = \frac{dv_t}{dt} = r\frac{d\omega}{dt} = r\alpha$$

5.2 Whole body rotational dynamics

In the previous example, a dancer experienced a rotational acceleration due to the presence of friction. For a dancer to initiate a turn, going from zero rotational velocity to spinning around a vertical axis, there must be some sort of net "rotational force" present. The rotational analog of force is torque, and the definition of torque is given by the vector cross product

$$\vec{\tau} = \vec{r} \times \vec{F} \tag{5.4}$$

where $\vec{r}$ is the vector extending from the axis of rotation to the point of application of the force, $\vec{F}$. Recall from Chapter 4 that the magnitude of the torque is $\|\vec{\tau}\| = rF \sin \phi = F_\perp r = Fr_\perp$ where ϕ is the angle between $\vec{r}$ and $\vec{F}$. Torque is greatest for a given $\vec{r}$ and $\vec{F}$ when the vectors are perpendicular to one another, and its magnitude can be found by either considering the component of $\vec{F}$ perpendicular to $\vec{r}$ ($F_\perp$) or the component of $\vec{r}$ perpendicular to the line of action of $\vec{F}$ ($r_\perp$). Recall also that the quantity $r_\perp$ is the *moment arm* for the force, and a larger moment arm for a given force increases the torque produced around an axis of rotation.

Imagine a stick on a horizontal, frictionless table that can rotate around a frictionless hinge at one end (Fig. 5.2). Three forces are being exerted on the stick. The torque due to F_1 is zero because the force acts at the axis of rotation ($\vec{r}_1 = 0$). The torque due to F_3 is also zero, since $\vec{F}_3$ and $\vec{r}_3$ are anti-parallel ($\sin \pi = 0$). The torque due to F_2 is $\tau_2 = r_2 F_2 \sin \phi$ in the clockwise direction (or into the page using the right-hand rule for vector cross product) (Fig. 5.3).

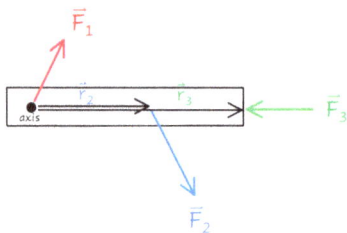

Figure 5.2: Three forces act on a rigid object at different locations and in different directions. The fixed axis of rotation of the object is perpendicular to the page and denoted by the point on the left.

Now imagine a dancer balancing at rest in passé relevé (the body position on the left in Fig. 5.1). That dancer would have a very difficult time generating the torque necessary to begin rotating on their own with only the ball of one foot in contact with the ground, but a partner standing behind the dancer could generate the torque necessary to spin her by holding onto either side of the dancer's waist, pushing forward with one hand, and pulling backward with the other. The partner must also be careful to exert forces of equal magnitude so that the net force on the dancer is zero. Otherwise, the dancer's CM would experience a translational acceleration away from the balanced position over her base of support.

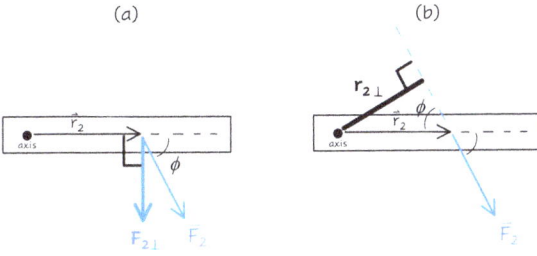

Figure 5.3: To find the magnitude of torque due to $\vec{F}_2$ around the axis of rotation, we can (a) find the component of $\vec{F}_2$ perpendicular to $\vec{r}_2$, and $\tau = F_{2\perp} r_2$, or (b) find the moment arm, $r_{2\perp}$, and $\tau = F_2 r_{2\perp}$.

Just as Newton's second law for translations states that a net force on a system causes a change in linear momentum, Newton's second law for rotations states that a net torque on a system causes a change in its rotational momentum, $\vec{L}$ (eq. (5.5)).

$$\sum \vec{\tau} = \frac{d\vec{L}}{dt} \tag{5.5}$$

Similar to the linear quantity $\vec{p} = m\vec{v}$, its rotational equivalent is

$$\vec{L} = \mathbf{I}\vec{\omega} \tag{5.6}$$

where $\mathbf{I}$ is rotational inertia. Conceptually, inertia quantifies the object's resistance to changes in rotational velocity (similar to mass and its resistance to changes in linear velocity). However, $\mathbf{I}$ is not a simple scalar quantity like mass, but a 3×3 tensor (matrix), as will be discussed in Section 5.8. For simple geometric shapes rotated around a fixed symmetry axis known as a principal axis (as is the case in introductory physics), rotational inertia *does* reduce to a scalar quantity known as the moment of inertia, and then $\vec{L}$ and $\vec{\omega}$ are in the same direction, but this is not the case in general. In this simplified case, in which I is a scalar, can we rewrite eq. (5.5) as $\Sigma\vec{\tau} = I\vec{\alpha}$, reminiscent of its linear counterpart, $\Sigma\vec{F} = m\vec{a}$.

Rotational inertia depends on the object's mass and how the mass is distributed around an axis of rotation. More mass distributed further from the axis of rotation leads to a larger rotational inertia. A dancer with the leg extended to the side (à la seconde) or posteriorly (arabesque) has a larger rotational inertia around a vertical axis than standard pirouette position or one with the legs down because the extended leg places more mass further from the vertical axis of rotation (Fig. 5.4).

For a point particle of mass m located a perpendicular distance r from the axis of rotation, $I = mr^2$. Total inertia of a system of many particles rotated around a symmetry axis (each with mass m_i and perpendicular distances r_i) can be found simply by summing all of their individual inertias, $I = \sum m_i r_i^2$, or for a continuous mass distribution, the sum becomes an integral, $I = \int r^2 \, dm$. Moments of inertia of various geometric shapes can be found in many introductory physics textbooks (e. g., a uniform sphere (mass M, radius R) rotated around an axis through its center ($I_{\text{sphere}} = \frac{2}{5}MR^2$), a uniform

Larger Inertia   Smaller Inertia

Figure 5.4: More mass further from the axis of rotation (dashed vertical line) leads to a larger moment of inertia for the dancer on the left. Her arms and gesture leg place mass further from the axis of rotation than the dancer on the right, whose arms and legs are aligned closer to the vertical axis of rotation.

cylinder or disk (mass M, radius R) rotated around its longitudinal axis ($I_{disk} = \frac{1}{2}MR^2$), etc.), and all have standard units of kg·m^2.

Another useful result from introductory physics is the parallel axis theorem, which relates the moment of inertia of an object (mass M) around an axis through its center of mass (I_{CM}) to the moment of inertia around another axis parallel to the first but that has been displaced a distance d (Fig. 5.5). The *parallel axis theorem* states

$$I = I_{CM} + Md^2 \tag{5.7}$$

For example, for the uniform cylinder in Fig. 5.5, if $d = R$, so that the cylinder is rotated around an axis tangent to its long edge, $I = I_{CM} + MR^2 = \frac{1}{2}MR^2 + MR^2 = \frac{3}{2}MR^2$.

axis

d

Figure 5.5: The parallel axis theorem can be used to find the moment of inertia of an object around an axis that has been displaced from and remains parallel to an axis through its CM.

This base of knowledge from introductory physics is useful, but must be expanded upon to study rotation of a complex object such as the human body. Dancers are not rigid geometric shapes only ever spun around a symmetry axis. The following sections discuss the different phases of whole-body dance rotations (push off for a turn, the turn phase itself, and landing of turns), applying physical principles of rotations and discussing biomechanical research done in these areas. Turns performed in the air are also discussed, as well as how to determine the whole body inertia tensor of a dancer. Dynamic balance of dance rotations is also covered, and finally, rotations of body segments around joints and 3D joint angle measurements.

5.3 Push off for a turn

5.3.1 Torque

During the preparation for a solo turn like a pirouette, a dancer exerts forces with the feet against the ground in order to produce torque around a vertical axis. We will use the standard en dehors ballet pirouette as our primary example, and the principles described here can be applied to any style of turn. To initiate a pirouette from fourth position as shown in Fig. 5.6, a dancer pushes with both feet against the ground, and the horizontal components of the ground reaction force at each foot (F_1 and F_2) produce torque around the vertical axis. (If we placed force plates beneath each foot, we could measure F_1 and F_2.) The torques generated by these forces ($\vec{r}_1 \times \vec{F}_1$ and $\vec{r}_2 \times \vec{F}_2$) are both clockwise from the top-down view in Fig. 5.6. If the pirouette were initiated instead from fifth position with the two feet touching, $\vec{r}_1$ and $\vec{r}_2$ would be smaller, so to produce the same magnitude of torque the dancer would need to increase F_1 and F_2 with a stronger push against the floor.

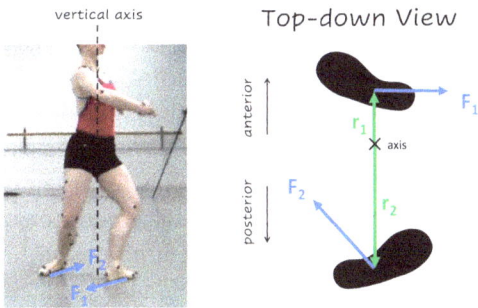

Figure 5.6: Horizontal ground reaction force exerted on each of a dancer's feet during pushoff for an en dehors pirouette (turn clockwise from the top-down view). Horizontal displacements of each point of force application from the vertical axis of rotation are shown in green in the top-down view. Vertical components of the GRFs are present but not displayed here.

Taking a wider fourth position effectively increases the moment arms for the torque; however, there are potential negative consequences one must consider if the stance is greatly widened. One such consequence is that it can place the dancer's CM further behind the front foot, which will become the supporting foot (BoS) of the pirouette. Not only does the dancer need to generate rotational momentum during pushoff, but linear momentum to launch the CM anteriorly toward a vertical position over the BoS [113]; otherwise, the turn will not be balanced. This is similar to the dancer in Chapter 4 needing to launch the CM anteriorly for a dynamically balanced piqué arabesque, but with an added whole body rotation. F_2 in Fig. 5.6 has an anterior component, meaning the back foot exerts force to push the CM forward. (It should also be noted that a slight medial-lateral net force toward the supporting foot is needed as the dancer rises up onto pointe or demi-pointe).

One study found that dancers were able to perform more controlled revolutions in a pirouette by increasing the distance between the feet during preparation, so long as 60 % of the dancer's body weight remained over the front foot [96]. That distribution of weight seemed to be ideal for effective torque generation and later balance control of the turn. Another study found that dancers who took a preparatory stance with the CM closer to the front foot generated more $\vec{L}$ with the back leg compared to dancers who stood with the CM displaced further from the front foot [113]. The results of that study suggested that there is not necessarily an optimal width for pirouette preparation, but strategies may be adapted depending on preparatory position. If there is not an optimal preparation width, the dancer having difficulty performing the triple turn may be encouraged to take a comfortable fourth position, or the width preferred by his dance teacher. Shifting his weight slightly toward the front foot and pushing somewhat more with the back foot may be a helpful strategy.

Now imagine a dancer wishes to begin a turn from rest, with only one foot on the ground during pushoff. The vertical axis of rotation already runs through the single supporting foot. Our intuition says it will be more difficult for the dancer to generate rotational momentum from the single foot preparation. It is still possible, however, to generate torque and initiate a whole body rotation by twisting the supporting foot against the ground. Let's assume the dancer's CM remains along the vertical axis of rotation during the single stance pushoff, so the CM acceleration is zero. If there was a force plate beneath the dancer's foot, it would measure zero net horizontal force. This doesn't mean that there is no horizontal force at all between the dancer's foot and the ground (there is still friction after all), it simply means that friction acts in different directions on different parts of the foot as it pushes against the ground, and if we add up all of these frictional force vectors, they would sum to zero. As the foot twists against the ground (not slipping, but remaining fixed in place), the static frictional forces are directed opposite to the impending rotational motion of the foot. If we imagine the small amount of friction acting at a point on the foot, it acts tangent to the dashed, circular line in Fig. 5.7. To find the total torque due to friction from the twisting of the foot, we would add up, or integrate, the contributions of all of the infinitesimal torques ($d\vec{\tau} = \vec{r} \times d\vec{F}$),

$$\vec{\tau} = \int \vec{r} \times d\vec{F} \tag{5.8}$$

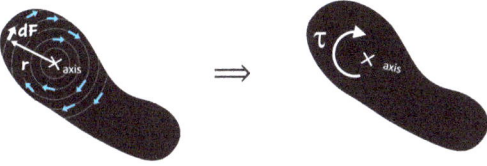

Figure 5.7: Free torque due to friction between the foot and the ground. The free torque can be due to static friction as in during the pushoff for a turn, or kinetic friction, as the foot rotates against the ground during the turn phase.

Top-down View

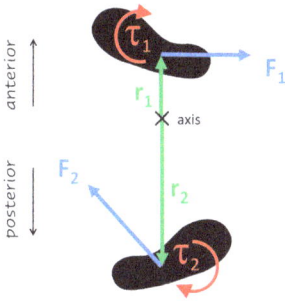

Figure 5.8: Top down view of the horizontal ground reaction forces and free torques between a dancer's feet and the ground during 4th position preparation for a pirouette.

This torque, due to the twisting of the foot against the ground, is referred to as the *free torque* and can be measured directly with a force plate.

In general during the pushoff for a whole body rotation from two feet, both the horizontal GRF and the free torque at each foot contribute to the net torque around a vertical axis (Fig. 5.8), and this net torque leads to a change in rotational momentum.

$$\Sigma \vec{\tau} = (\vec{r}_1 \times \vec{F}_1) + (\vec{r}_2 \times \vec{F}_2) + \vec{\tau}_1 + \vec{\tau}_2 = \frac{d\vec{L}}{dt} \tag{5.9}$$

The net force leads to a change in linear momentum of the CM

$$\Sigma \vec{F} = \vec{F}_1 + \vec{F}_2 + \vec{F}_g = \frac{d\vec{p}}{dt} \tag{5.10}$$

where $\vec{F}_1$ and $\vec{F}_2$ are the 3D *GRF* vectors (including both horizontal and vertical components) and $\vec{F}_g$ is the force of gravity. Eq. (5.9) and eq. (5.10) can be modified for any number of supporting limbs during pushoff, whether it's a single legged pushoff or if the pushoff includes forces from one or both hands (e. g., for a headspin). Different styles of turns have different mechanical objectives, so will require different amounts of rotational and linear momentum to be produced from the pushoff. For example, there is a difference between stationary turns like the pirouette, in which the CM is not meant to translate horizontally once the dancer reaches the turn phase, and traveling turns like piqué turns, where the CM undergoes purposeful horizontal displacement (Fig. 5.9). In the piqué turn, the CM begins further from the BoS when the gesture leg leaves the ground compared to pirouettes, as measured by the initial topple angle ($\theta_0 \approx 20°$ for piqué compared to $\theta_0 \approx 10°$ for pirouettes, as shown in Fig. 5.10). Pirouette turns were also measured to take a longer duration of time than piqué turns with the same number of revolutions (single or double) [114].

Figure 5.9: Ballet pirouette and pique turns. Pique turns are initiated with a single supporting limb during pushoff (that becomes the GL for the turn). The pique turn is a traveling turn, in which the CM translates (in the y-direction in this figure), whereas in an ideal pirouette, the CM has zero translation during the turn phase. The body position with the GL in passé is very similar between the two turns. Reprinted with permission from [114].

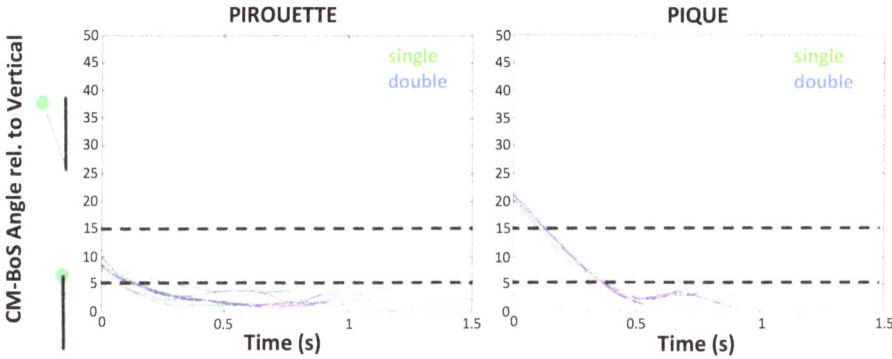

Figure 5.10: Topple angle as a function of time for single turn (green) and double turn pirouttes (left) and pique turns (right). Time $t = 0$ corresponds to when the GL foot leaves the ground. Reprinted with permission from [114].

It takes quite a bit of practice for novice dancers to learn how to exert these forces with the feet against the ground to push off for a successful pirouette. Dancers must exert enough force with the feet on the ground to generate the rotational momentum needed while simultaneously scaling these forces in just the right way to launch the CM over the supporting foot. If the anterior component of F_2 is too large or small or if F_1 and F_2 do not have similar magnitude of force in the medial-lateral direction, the pirouette will be off balance. While a dancer need not launch the CM to a perfectly balanced position for an ultimately successful turn, if the initial deviation from balance is too great, drastic measures like hopping on the supporting foot or landing the turn early to prevent a fall will become necessary.

5.3.2 Rotational impulse and motor control strategies

Optimizing the rotational momentum of a pirouette or a given style of turn does not necessarily mean maximizing it. Ultimately dancers and choreographers are more con-cerned with factors such as the aesthetics of the turn and that it is balanced, the turn's rotational velocity (e. g., so the movement remains in tempo with music), or the total number of revolutions completed before ending the turn. Rotational velocity is affected by both $\vec{L}$ and inertia, but there are often technique or choreographic constraints on body configuration for a turn. If the turn is in a position with large rotational inertia (e. g., arabesque) and a dancer finds it difficult to achieve the desired number of rev-olutions or $\vec{\omega}$, a greater $\vec{L}$ is needed. Studies have also shown that $\vec{L}$ generated during pushoff does increase as the number of turns increases in both pirouettes and piqué turns [47, 113, 112]. If we think back to the dancer at the beginning of the chapter who wished to achieve a triple turn pirouette, this seems to indicate more rotational momen-tum would help, but what mechanisms and strategies might a dancer effectively use to increase $\vec{L}$ during pushoff?

As done with translational motion, we can rewrite Newton's second law for rotations (eq. (5.5)) into a *rotational impulse-momentum theorem*.

$$\int_{t_0}^{t} \sum \vec{\tau} \, dt = \Delta \vec{L} \tag{5.11}$$

where the quantity $\int \Sigma \vec{\tau} \, dt$ is the rotational impulse. This means that a greater change in rotational momentum can be achieved by exerting a larger net torque and/or over a longer duration of time ($\Delta t = t - t_0$). These two options may make the answer seem simple, but when we begin to consider a dancer's options for how to go about increasing the rotational impulse, we see the answer is not so straightforward. Applying this to pirouettes (or any turn initiated from a double supported pushoff), the dancer can extend the duration of the pushoff phase and/or increase the net torque around the vertical axis due to the horizontal ground reaction forces and/or free torques at the two feet.

$$\int_{t_0}^{t} (\vec{r}_1 \times \vec{F}_1 + \vec{r}_2 \times \vec{F}_2 + \vec{\tau}_1 + \vec{\tau}_2) \, dt = \Delta \vec{L}$$

Something that limits the amount of time available for a dancer to exert torque with the feet is that as $\vec{L}$ increases, the body begins to rotate and then the feet must follow! Once the dancer's trunk rotates through roughly half of a revolution, the gesture leg foot leaves the ground and the feet can no longer actively generate $\vec{L}$. A strategy identified to extend the amount of time that the dancer's trunk and legs can maintain their forward facing position is by storing momentum in the arms [56]. If a dancer rotates extended arms in the direction of the turn during push off, the arms can take on some of the generated $\vec{L}$ during their windup. With the arms extended far from the axis of rotation, their larger inertia allows rotational velocity to be relatively lower for a given $\vec{L}$. On the other hand, if the arms did not move independently of the rest of the body, $\vec{L}$ generated with the pushoff would cause the dancer's whole body to begin to rotate immediately. One study experimentally measured this storing of $\vec{L}$ in the arms and found that the lead arm and the trail arm carry the two greatest amounts of $\vec{L}$ of any body part during the pushoff, and that momentum was then transferred to other parts of the body during the turn phase [47]. Interestingly, however, that same study found that although the windup was more pronounced and the $\vec{L}$ generated increased as the number of turns in a pirouette increased, duration of time of the pushoff remained the same. This suggests that the total generated rotational momentum increased via larger force(s) and/or torques exerted with the feet against the ground. Perhaps the larger net torque would have otherwise led to a lesser total pushoff time if not for the increased windup. Force data were not reported in this particular study, however, so the specifics of what changes in ground reaction forces and/or free torques may have been made were unknown.

There are multiple mechanical ways to modify net torque during pushoff, including changing the magnitudes of the free torques, the magnitudes or directions of $\vec{F}_1$ or $\vec{F}_2$ or their moment arms, or any combination of those options. We should also remember that these ground reaction forces vary with time, so a dancer could modify the force profiles during pushoff. One study found that when going from single to double turn ballet pirouettes, the main strategy used by dancers to increase $\vec{L}$ was to increase the net horizontal GRF at the front and/or back foot ($\vec{F}_1$ and/or $\vec{F}_2$), but that strategies varied between dancers [113]. In this case, a dancer might scale the same muscles used in a single turn but with a greater force to generate greater ground reaction force and torque. If a strategy is used instead to apply the GRF in a different direction to achieve greater $\vec{L}$, this could require "different muscle recruitment timing or muscle sets" [112]. It may help the dancer at the beginning of the chapter to adopt the simpler strategy of trying to push in the same manner and with the same muscles, but exert slightly larger forces and use a bit more arm windup. The dancer must be careful, however, to not make the windup too pronounced or it may negatively impact the aesthetic of the pirouette.

Another study compared control strategies used during pushoff for single pirouettes with ballet turnout or from parallel position (zero hip external rotation) [115]. Several differences, but also similarities, were found between strategies with the legs oriented in the two different positions. There were subject-specific differences in hip and knee muscle activation patterns (measured with electromyography, which will be discussed in Chapter 6) as well as joint kinetics. In spite of differences at the individual muscle or joint level, it was found that $\vec{F}_1$ and $\vec{F}_2$ were mostly aligned with their respective leg plane (the plane made by the thigh and shank segments) during pushoff for both types of pirouette. Additionally, for both parallel and turnout positions, the net joint torques for each leg were primarily flexion/extension torques. During pushoff for both types, most dancers utilized hip extensor and abductor torques at the front leg (which becomes the supporting leg during the turn), and ankle plantarflexor, knee extensor, hip flexor and abductor torques at the back leg (which becomes the gesture leg) [115].

One of the takeaways from research on pushoff for turns is that strategies to increase $\vec{L}$ tend to vary between dancers, but these different strategies can lead to similar overall $\vec{L}$ generated and ultimate kinematics of the turn. Turns performed by different dancers can look the same to an outside observer, but which muscles were recruited and how the ground reaction forces were regulated may have been different. Because different dancers have been found to use different motor control strategies to increase $\vec{L}$ and achieve the same kinematic results, researchers suggest individualized training based on a dancer's specific chosen technique [112]. This makes it more difficult to come up with general tips for any given dancer attempting to improve their pirouettes; however, dancers and dance educators equipped with biomechanical knowledge can attempt educated technique modifications as opposed to uninformed trial-and-error attempts such as those being made by the dancer at the beginning of the chapter. As a whole this may lead to more efficient learning as well as reduce the risk of overuse injury by many futile attempts.

5.4 Turn phase

Once the pushoff (or turn initiation) phase ends and the turn phase begins (Fig. 5.9 and Fig. 5.11), one might incorrectly assume that after a successful pushoff a dancer can hold a fixed body position and execute the turn like a passively spinning top. It is true that for turns like the piqué and pirouette, $\vec{L}$ is no longer actively generated by the dancer but decreases due to a frictional torque. This is unlike turns such as fouettés, in which the dancer generates more rotational momentum with the supporting foot by coming down into plié once per revolution. The dancer's body position may also not change substantially (or noticeably) during the turn phase (e. g., maintaining passé position), which can lead to the impression that a dancer holds the body rigidly, other than spotting of the head, until landing the turn. If we look closely at Fig. 5.11, we can make a few qualitative observations about how the shape of this dancer's passé position has changed from the start (left most picture) to the end (right most picture) of the rotation. The dancer's hands are further apart and her rib cage looks to be shifted further toward her gesture leg at the beginning, compared to her arms being brought in and what looks to be an improved alignment by the end of the rotation. These types of subtle differences can be challenging to observe in real time as a dancer rotates.

Figure 5.11: One revolution during the turn phase of a pirouette, with time evolving from left to right. This is the second full turn of a triple turn pirouette.

Research has shown that successfully executed pirouettes have a much more dynamic turn phase than meets the eye. It has been demonstrated that to perform more than three revolutions in a pirouette with typical initial rotation rates and topple angles, a dancer must make active adjustments for balance maintenance while rotating in the turn phase [64]. Otherwise, the dancer would topple to the ground, much like the inverted pendulum studied in Chapter 4. Before discussing the dynamic nature of the turn phase, let's first gather some illuminating information by approximating the dancer as a rigid body like the spinning top.

5.4.1 Slowing frictional torque

We can use rigid body dynamics to estimate how many revolutions in a pirouette might be possible from a single pushoff if balance maintenance were not an issue. Is it eleven revolutions, the number famously executed by Michael Barishnikov in the 1985 movie *White Nights*? Or is it more? The exact number will vary depending on several factors including the coefficient of friction between the dancer's shoes and floor and the initial rate of rotation. We will solve for $\Delta\theta$ before coming to rest symbolically in terms of inertia of the dancer around the spin axis (I), initial rotational velocity (ω_0), coefficient of friction (μ_k), mass of the dancer (M), and BoS radius (R). Then the numerical estimate for $\Delta\theta$ can be computed for a variety of scenarios.

Example. Estimate the number of revolutions a dancer can perform in a pirouette before friction slows the turn to a stop. We make the following simplifying assumptions for this problem: 1) the dancer maintains balance with the CM over the BoS during the turn, 2) vertical GRF (normal force, F_N) is equal to the dancer's body weight (Mg) for all times, 3) pressure is uniformly distributed over the area of support (A) between the foot and floor ($P = \frac{dF_N}{dA} = \frac{Mg}{A}$), 4) the area of support is a circle with radius R, and 5) air resistance is negligible.

Solution. When the dancer performs a pirouette with rotational velocity around the vertical axis ($\hat{z}$), $\vec{\omega} = \omega\hat{z}$, the kinetic friction force will exert a torque in the opposite direction of the pirouette, $\vec{\tau} = -\tau\hat{z}$, slowing the turn to a stop.

For an area element dA of the contact area of dancer's foot, the kinetic friction force opposes the instantaneous velocity of that area element, so $d\vec{F}_{f,k} \parallel -\vec{v} \perp \vec{r}$. From $d\vec{\tau} = \vec{r} \times d\vec{F}_{f,k}$, the magnitude of the torque that's applied to an area element dA is

$$d\tau = r\mu_k dF_N.$$

We use the definition of pressure to write $dF_N = PdA$, where $dA = rdrd\theta$ in polar coordinates, and since the pressure is uniformly distributed, $P = \frac{Mg}{A} = $ constant. Making this substitution for dF_N, we can find the total free torque due to friction (eq. (5.8)).

$$d\tau = r\mu_k dF_N = r\mu_k \frac{Mg}{A} rdrd\theta$$

$$\tau = \frac{\mu_k Mg}{A} \int_{\theta=0}^{2\pi} \int_{r=0}^{R} r^2 \, dr \, d\theta = \frac{\mu_k Mg}{A}\left(\frac{R^3}{3}\right)(2\pi)$$

Since we are approximating the BoS as a circle, $A = \pi R^2$, and the magnitude of the total free torque due to kinetic friction is

$$\tau = \frac{2}{3}\mu_k MgR$$

so

$$\vec{\tau} = -\frac{2}{3}\mu_k MgR\hat{z}$$

Now that we have an expression for $\vec{\tau}$, we can find the rotational acceleration (which will be constant since τ is constant), and then use kinematics to find $\Delta\theta$ and the number of revolutions before stopping.

$$\Sigma\vec{\tau} = I\vec{\alpha} \quad \longrightarrow \quad \vec{\alpha} = \frac{\Sigma\vec{\tau}}{I} = -\frac{2\mu_k MgR}{3I}\hat{z}$$

Then,

$$\omega^2 = \omega_0^2 + 2\alpha\Delta\theta$$

$$0 = \omega_0^2 + 2\alpha\Delta\theta$$

$$\longrightarrow \Delta\theta = \frac{-\omega_0^2}{2\alpha} = \left(\frac{-\omega_0^2}{2}\right)\left(\frac{-3I}{2\mu_k MgR}\right)$$

$$\Delta\theta = \frac{3I\omega_0^2}{4\mu_k MgR}$$

We will assume values for μ_k, M, R, I and ω_0 that are reasonable estimates for a dancer performing a standard ballet pirouette on demi pointe in canvas ballet slippers ($\mu_k = 0.1$, $M = 55$ kg, $R = 2.5$ cm, $I = 1.5$ kg·m^2, and $\omega_0 = 2.5$ rev/s = 5π rad/s). (The method to determine whole body inertia is discussed in Section 5.8.2.) Numerical values can be changed for different styles of turn (I), different shoes or flooring (μ_k), etc. We could also choose a different pressure distribution instead of the assumed uniform distribution in our analysis.

$$\Delta\theta = \frac{3I\omega_0^2}{4\mu_k MgR} = \frac{3(1.5 \text{ kg·m}^2)(5\pi \text{ rad/s})^2}{4(0.1)(55 \text{ kg})(9.8 \text{ m/s}^2)(0.025 \text{ m})}$$

$$\Delta\theta = 205 \text{ rad}\left(\frac{1 \text{ rev}}{2\pi \text{ rad}}\right) = 32 \text{ rev}$$

Thirty-two revolutions is a very large number for a pirouette. While this is only an estimate, it is a good indication that in practice, friction is not the main factor limiting the number of turns a dancer can complete in a standard ballet pirouette.

5.4.2 Balance maintenance during rotations

The pirouette's "turn phase," as depicted in Fig. 5.9, actually contains two subphases: the ascent and the turn. During the ascent the gesture leg moves from the ground to its passé position at the knee. Also during ascent, the CM is moving toward its (hopefully balanced) position over the BoS. This movement of the CM toward the BoS and motion of the GL make the ascent phase distinct from the rest of turn phase from both a mechanical and motor control perspective. For the remainder of the pirouette's turn phase, there should be little CM displacement while the pirouette remains balanced and passé position is maintained.

Research has found that when a pirouette was a double instead of a single turn, dancers adjusted the GRF relative to the CM during the pirouette's ascent. For the double turns, dancers utilized larger braking forces and torques to prevent the CM from overshooting the BoS [114]. This method of increasing braking forces and torques during ascent was also observed in double turn piqués compared to singles, serving to decrease horizontal velocity as the CM approached vertical alignment over the BoS [114].

Like statically held poses discussed in Chapter 4, balance maintenance of the pirouette involves adjusting the relative positioning of the CM and BoS. Unlike a statically held pose, however, the BoS is rotating against the ground with relative sliding motion between the foot and support surface. The dancer has three options: (1) accelerate the CM back toward a position over the BoS, (2) accelerate the BoS back under the CM, or (3) a combination of CM and BoS acceleration. The physical mechanisms that enable a dancer to carry out the above strategies are (1) Exert a horizontal force with the BoS against the floor toward the direction of topple (so that the GRF is away from topple), (2) hop the BoS to a position vertically in line with the CM (aesthetically undesirable), or (3) decrease the frictional force at the floor (by decreasing vertical GRF) so that the BoS can be slid to a position under the CM.

During a pirouette, the relative sliding between the BoS and the floor as the bottom of the foot rotates against the ground is an example of kinetic friction. Balancing at rest in the static pirouette position, there would be static friction between the foot and ground. An adjustment (e. g., a hip strategy) that would have led to CM acceleration and no BoS movement in the static case could instead lead to a BoS translation in the rotating case since kinetic friction is less than static friction. The more subtle sliding of the foot will lead to movement of both the BoS and CM and is a combination of the physical mechanisms in (1) and (2). When sliding the BoS toward a position under the CM, friction exerts a force on the BoS in the opposite direction. The force of friction will cause the CM to have an acceleration away from the direction of topple (Fig. 5.12).

A study investigating correlations between number of revolutions (n) in a pirouette (ranging from 1 to 5) to various parameters found a significant positive correlation ($r = .873, p < 0.001$) between n and BoS distance traveled per revolution (normalized to dancer's height) [63]. Number of revolutions was not correlated with any of the other parameters measured (initial topple angle from vertical, margin of stability, spin rate, maximum topple angle, and dancer effective pendulum length). These results demonstrated that these dancers did not achieve more revolutions in pirouettes by beginning them closer to balance, rotating faster, or tolerating larger topple angles. The greater

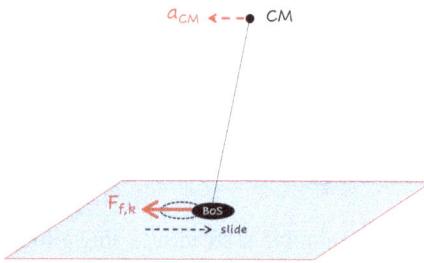

Figure 5.12: If the supporting foot (BoS) is slid toward the direction of topple (to the right in this figure), the force of kinetic friction ($F_{f,k}$) on the dancer is directed opposite of topple (to the left). This external force results in an acceleration of the dancer's center of mass (a_{CM}) back toward balance (the left).

number of revolutions were, however, accompanied with greater translation of the BoS per turn. This suggests that subtle sliding of the supporting foot can be an effective mechanism for balance maintenance during pirouettes. Dancers must be careful of course to make adjustments subtle enough so that the BoS translations are barely noticeable and do not become large hops.

If dancers are making adjustments during pirouettes, what balance maintenance strategies are being utilized by dancers and are they similar to strategies used during standing balance (e. g., ankle and/or hip strategies discussed in Chapter 4)? One study investigating joint angle kinematic coordination strategies during pirouettes discovered that successfully executed pirouettes exhibited a combined ankle-hip coordination, although specific coordination patterns varied between individuals and across trials [65]. Results highlighted the importance of the ankle joint with group mean ankle ab/adduction excursion significantly greater than all other joint angle excursions. Researchers also found evidence that supporting limb and pelvis/trunk joint angles oscillated at the pirouette's rotational frequency (one cycle per revolution), indicating that the coordination was continuously executed throughout the pirouette and in sync with the rotation. Like balance maintenance during non-rotating postures, the body's joints should not be held rigidly during rotations so that appropriate adjustments for imbalance can be made.

Actively applying adjustment mechanisms during pirouettes depends on a dancer's ability to sense an imbalance and make adjustments with the correct magnitude, direction, and timing. The rotation of the body complicates matters in a variety of ways. For one thing, as the body topples in a fixed direction relative to the stage frame, that direction continuously changes relative to the dancer's rotating body frame. For example, to an audience member, a dancer may topple to the left, but as the dancer rotates, their relative topple direction changes from anteriorly to laterally to posteriorly, etc. This may make it especially difficult to activate the appropriate muscles to execute a strategy for balance maintenance. Additionally, sensory information that a dancer relies on for balance is likely impacted and may be less effective during rotations. Proprioceptive information about subtle deviations of the body or its parts from equilibrium may be diluted by the overall sense of the rotational movement of the body. As the body and the head rotate, the dancer's field of vision rotates, and the motion of the head affects the fluid in the inner ear, which impacts vestibular feedback. Spotting the head may be a useful mechanism to provide brief, non-rotating visual feedback to a dancer, but much remains to be discovered about the utility of spotting while turning. There are still things to learn in general about balance during rotations, including whether different muscle synergies exist for balance during rotations, and whether expert turners utilize different muscle synergies than non-experts.

5.5 Spotting

Spotting the head is a commonly used technique for many types of whole body rotations in dance. Spotting refers to a dancer holding the head in a fixed direction once per turn (often toward the audience) while the rest of the body rotates, and then rapidly rotating the head around to the initial orientation. Greater isolation of the head and keeping the gaze on a fixed spot for as long as possible have been identified by dance experts as criteria important for successful spotting technique [30]. Artistically, movements in dance are performed with the audience in mind, so keeping the gaze toward the audience during a turn has a certain aesthetic appeal. Some dance teachers assert that vigorously spotting the head can force an extra revolution in a multiple turn pirouette [30]. A dancer cannot gain angular momentum in a turn, however, simply by rotating the head. The angular speed of the rest of the body will in fact slow as the head absorbs some of the angular momentum when it rotates during spotting. To minimize this slowing effect, the head should be kept on the axis of rotation, where its moment of inertia is smallest [55].

If spotting is not useful to help a dancer gain more rotational momentum, what may be its functional utility? As previously mentioned, it may help dancers gather visual feedback useful for balance maintenance. Dance experts have also suggested it may promote the rhythm of the turn, reduce dizziness, and assist with dancer's orientation [30]. Spotting appears to increase in importance when performing multiple turns in a pirouette and as a dancer's pirouette skill level increases. Spotting was found not to be as useful when first learning pirouettes. One study compared novices separated into an experimental group (pirouette training with spotting instruction) and a control group (pirouette training without spotting instruction) [48]. After eight weeks of training, few in the experimental group even demonstrated the spotting technique, and the experimental group did not perform better than the control group in double or multiple turn pirouettes. Balanced, whole-body revolutions are a very complex motor skill to learn, so learning head isolation in addition to whole body rotation seems to be too complicated to be implemented early in skill acquisition. This is useful information for dance educators, so spotting technique can be emphasized later in pirouette skill development. The dancer who is already practiced at double pirouettes and is attempting triples could however benefit from working on his spot. The dancer should choose a stationary reference point (e. g., a mark on the wall) to focus his eyes on once per turn. This could help orient him and provide visual information for sensory feedback, even though it can not gain him any extra momentum for the turn.

5.6 Landing and transitioning out of turns

Once a dancer completes the desired number of rotations in a turn, they must either land the turn in a static position or transition smoothly into the next movement. Landing or transitioning out of rotations is the least studied phase of dance turns, even though a poorly executed ending detracts from the overall quality of the turn. In the case in which a dancer lands the turn in a static position (e. g., with a plié in fifth position), for an optimal landing the dancer's supporting foot does not move to a different position on the ground. The heel should simply come down to the ground using ankle dorsiflexion and without any noticeable translation of the supporting foot. To accomplish this, the CM must be relatively close to a vertical position over the BoS at the end of the turn for a fifth position landing.

For the dancer to land the turn at rest, their rotational momentum must be ceased. Friction may have been decreasing the dancer's momentum continuously during the turn, but if a dancer still has a large rotational velocity at the time they wish to land, it can be difficult to do so in a controlled manner. One method to slow the dancer's rotation without even changing the total rotational momentum is for the dancer to increase their inertia by extending the arms and/or leg right before landing. Dancers often extend the arms outward during landing from pirouettes which not only decreases the rate of rotation around the spin axis but helps a dancer maintain balance with an increased inertia around the topple axes (similar to a tightrope walker holding a long pole). Once the feet come back into full contact with the ground during landing, the horizontal GRFs and free torques at each foot produce a larger negative rotational impulse, to stop the rotational momentum.

If a dancer transitions smoothly to another movement immediately following the rotation instead of landing at rest on two feet, they must anticipate where their body needs to go next and adjust accordingly. The dancer's CM will not be in the same position every time at the end of a pirouette. Their CM may off balance to the left and need to be projected to the right to travel out of the turn. A dancer will need to use feedforward control to modify how the movement after the pirouette is executed from differing initial conditions.

5.7 Turns in the air and conservation of rotational momentum

Whole body rotations in dance are not confined to cases where the dancer is in contact with the ground. Rotations of the body in the air can occur around a longitudinal axis

(e. g., ballet tour en l'air), anterioposterior axis (e. g., side aerial), mediolateral axis (e. g., front or back flip) or an axis that is not aligned with one of the body's anatomical axes. No matter the rotational axis, once the body is in the air, the only external force acting on the dancer is that due to gravity (ignoring a negligible amount of air resistance). $\vec{F}_g$ acts at the body's CM, so it can not produce a torque around an axis through the CM to lead to whole body rotational acceleration once the dancer is in the air. So, any rotational impulse around the desired axis of rotation must be generated during pushoff with the feet in contact with the ground.

No matter the type of turn in the air, both a rotational impulse and a linear impulse are generated during pushoff. Some rotating leaps may travel horizontally and vertically like the side aerial, ballet tour jeté, or Irish dance zooms, whereas others, like the tour en l'air, have mostly vertical CM motion. No matter the case, the same basic principles of the linear and rotational impulse-momentum theorems still apply (eq. (3.13) and eq. (5.11)). How much horizontal, vertical, and rotational momentum are needed to meet the mechanical demands of the task depend on the specific turning leap.

Once in the air, there is zero net torque on the dancer, so from eq. (5.5),

$$\text{If } \Sigma\vec{\tau} = 0, \quad \text{then } \frac{d\vec{L}}{dt} = 0 \tag{5.12}$$

which means the total rotational momentum of that system remains constant. This is a statement of *conservation of rotational momentum*. In other words, if we measure the total rotational momentum initially, it will be equal to the final rotational momentum at some later time so long as $\Sigma\vec{\tau} = 0$. A consequence of the conservation of rotational momentum is that for our system of one dancer, a change body configuration (I) leads to a change in rotational velocity (ω).

$$\vec{L}_i = \vec{L}_f$$
$$I_i\omega_i = I_f\omega_f \tag{5.13}$$

So if the inertia is reduced by bringing mass closer to the axis of rotation ($I_f < I_i$), then the dancer will rotate more rapidly ($\omega_f > \omega_i$), and vice versa, to keep $\vec{L}$ constant. This is sometimes referred to as the "ice skater effect." A breaker doing a headspin wearing a beanie (so there is very little friction between the head and floor) is much like the skater spinning on low friction ice where $\Sigma\vec{\tau} \approx 0$ around the spin axis. Holding the arms and legs out during pushoff with the hands for the spin allows for more time for the pushoff torque to be exerted. Then, if the arms and legs are brought in (to an upside down anatomical position or pencil position), the breaker can rotate extremely rapidly at the end of the headspin. In a turning leap such as the ballet tour jeté, dancers can create the illusion of leaping into the air with no rotation and then spinning seemingly out of nowhere. This effect is best achieved if the turn is begun with a large inertia (and ω is small enough to be nearly undetectable) and then the inertia is reduced for a larger, and now observable, rotational velocity [57].

Example. A dancer leaves the ground for a turning leap (e. g., tour jeté) in which they will complete some number of revolutions (or fraction thereof) around a vertical axis during the total time that their feet are off of the ground ($t = 0.5\,$s). They begin the turning leap with their arms and legs extended (inertia = I_0) and a rotational velocity of $\omega_0 = 0.4\,$rev/s for the first 0.1 sec. Then they bring their arms and legs closer to the axis of rotation (new inertia I) for the middle portion of the leap (lasting for $t = 0.3\,$s). Finally, they extend the arms and legs out once again (inertia = I_0) for the final 0.1 s of the leap. (a) What must be the ratio of $\frac{I_0}{I}$ such that the dancer completes $\frac{1}{2}$ rev during the turning leap? (b) If instead the dancer wishes to complete one full revolution, what must be the ratio of $\frac{I_0}{I}$, assuming all else remains the same?

Solution. Once the dancer's feet leave the ground, rotational momentum is conserved (eq. (5.13)).

$$I_0\omega_0 = I\omega$$
$$\frac{I_0}{I} = \frac{\omega}{\omega_0}$$

We know conceptually that when the dancer's arms and legs are closer to the axis of rotation, $I < I_0$ so $\omega > \omega_0$. In order to determine the precise ratio of $\frac{I_0}{I}$, we need to know the value of ω required to complete the necessary turns in the air. To do so, let's break the turning leap into the beginning (I_0, ω_0, $\Delta t = 0.1\,$s), middle (I, ω, $\Delta t = 0.3\,$s), and end (I_0, ω_0, $\Delta t = 0.1\,$s). During the beginning and end portions of the leap, the dancer completes 0.04 rev.

$$\Delta\theta_{\text{beg}} = \Delta\theta_{\text{end}} = \omega_0\Delta t = (0.4\,\text{rev/s})(0.1\,\text{s}) = 0.04\,\text{rev}$$

This means that the dancer must rotate 0.42 rev during the middle portion to complete the 0.5 rev total during the leap.

$$\Delta\theta_{\text{total}} = \Delta\theta_{\text{beg}} + \Delta\theta_{\text{mid}} + \Delta\theta_{\text{end}}$$
$$0.5\,\text{rev} = 0.04\,\text{rev} + \Delta\theta_{\text{mid}} + 0.04\,\text{rev}$$
$$\Delta\theta_{\text{mid}} = 0.42\,\text{rev}$$

To complete 0.42 rev in 0.3 s, requires a rotational velocity of

$$\omega = \frac{\Delta\theta_{\text{mid}}}{\Delta t_{\text{mid}}} = \frac{0.42\,\text{rev}}{0.3\,\text{s}} = 1.4\,\text{rev/s}$$

Therefore, the ratio of inertias is

$$\frac{I_0}{I} = \frac{\omega}{\omega_0} = \frac{1.4\,\text{rev/s}}{0.4\,\text{rev/s}} = 3.5$$

The inertia of the dancer's body with the arms and legs extended must be 3.5 times greater than that of the body with the arms and the legs pulled in.

(b) An identical approach is taken to find the ratio of inertias if the dancer completes a full revolution instead of half of a revolution, with the difference being that $\Delta\theta_{mid} = 1 - 2(0.04\,\text{rev}) = 0.92\,\text{rev}$, which requires $\omega = \frac{0.92\,\text{rev}}{0.3\,\text{s}} = 3.07\,\text{rev/s}$. This means $\frac{I_0}{I} = \frac{3.07\,\text{rev/s}}{0.4\,\text{rev/s}} = 7.7$. The ratio of I to I_0 is limited by the structure of the human body, so this large of a reduction in inertia may not be achievable in reality. The dancer would instead need to generate greater rotational momentum during pushoff to complete the full revolution.

It is also interesting to note that while rotational momentum is conserved in this example, the dancer's rotational kinetic energy is not. (Energy will be discussed in detail in Chapter 7.) The dancer's initial kinetic energy is $K_0 = \frac{1}{2}I_0\omega_0^2$ and is finally $K = \frac{1}{2}I\omega^2$. We make the substitution that $\omega = \frac{I_0}{I}\omega_0$ and have $K = \frac{1}{2}I(\frac{I_0}{I}\omega_0)^2 = \frac{I_0}{I}(\frac{1}{2}I_0\omega_0^2) = \frac{I_0}{I}K_0$. This means that for the dancer in part (a), the kinetic energy increased by a factor of 3.5 when the limbs were brought in and rotational velocity went up. But where did this energy come from? The muscles did work to bring the arms and legs in and as a result the dancer's total kinetic energy increased.

5.8 Inertia tensor

The remaining sections of this chapter provide more advanced information necessary for a comprehensive analysis of 3D rotations. These sections may be skipped without loss of continuity in an introductory biomechanics course.

So far in this chapter we still have not addressed the fact that the shape of the dancer's body makes the inertia more complicated than a simple geometric object. To get a more complete picture of rotational inertia, let's begin by revisiting our simple inverted pendulum (IP) model of a dancer (point mass on the end of a rigid rod). The IP can be used as a simple model for a rotating dancer. Imagine that the CM is *not* along a vertical line directly above the center of pressure as the dancer spins, so the IP makes a non-zero angle θ with the vertical (Fig. 5.13). As the IP rotates around the vertical axis, it sweeps out a cone shape, with $\vec{\omega} = \omega\hat{z}$ and the point mass moving in the horizontal plane. The rotational momentum of a point particle is

$$\vec{L} = \vec{r} \times \vec{p} \tag{5.14}$$

and the position of the point mass is $\vec{r} = x\hat{x} + y\hat{y} + z\hat{z}$ with linear momentum $\vec{p} = m\vec{v} = m(\dot{x}\hat{x} + \dot{y}\hat{y})$. We can relate the linear velocity of the point mass to its rotational velocity ω using eq. (5.3). Consider the top-down view in Fig. 5.13, where ρ is the distance of the point mass from the z-axis ($\rho = \sqrt{x^2 + y^2}$). We know that the magnitude of the tangential velocity, $v = \rho\omega$. The x- and y-components of $\vec{v}$ are

$$v_x = \dot{x} = -v\sin\phi = -(\rho\omega)\left(\frac{y}{\rho}\right) = -\omega y$$

$$v_y = \dot{y} = v \cos \phi = (\rho\omega)\left(\frac{x}{\rho}\right) = \omega x$$

Therefore, $\vec{p} = m\omega(-y\hat{x} + x\hat{y})$.

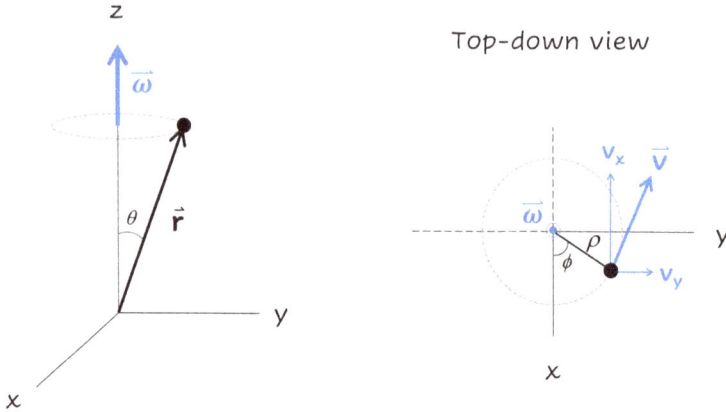

z

$\vec{\omega}$

Top-down view

θ $\vec{r}$

v_x $\vec{v}$

$\vec{\omega}$

ρ

ϕ

v_y

y

x

y

x

Figure 5.13: Simple inverted pendulum model of a dancer rotating around a fixed, vertical axis. (a) The IP makes a constant angle θ with the vertical (topple angle from Chapter 4) as the dancer rotates around z. (b) The top-down view of horizontal plane motion shows the rotational position of the IP, ϕ. The dancer's rotational velocity is $\vec{\omega} = \dot{\vec{\phi}}$, and the tangential velocity vector $\vec{v}$ as the point mass rotates around z is shown.

In general, the velocity of a point at position $\vec{r}$ on a rigid rotating body (rotational velocity $\vec{\omega}$) can be found using the vector cross product:

$$\vec{v} = \vec{\omega} \times \vec{r} \tag{5.15}$$

We are now equipped to determine the rotational momentum of the inverted pendulum, by applying the vector cross product in component form ($\vec{A} \times \vec{B} = (A_y B_z - A_z B_y)\hat{x} + (A_z B_x - A_x B_z)\hat{y} + (A_x B_y - A_y B_x)\hat{z}$):

$$\vec{L} = \vec{r} \times \vec{p} = m\omega\left[-zx\hat{x} - zy\hat{y} + \left(x^2 + y^2\right)\hat{z}\right]$$
$$= \left[(-mxz)\hat{x} + (-myz)\hat{y} + m\left(x^2 + y^2\right)\hat{z}\right]\omega$$

Let's first consider the z-component of $\vec{L}$, $L_z = m(x^2 + y^2)\omega = m\rho^2\omega$. The quantity $m\rho^2$ looks like the familiar moment of inertia for a point particle rotated around the z-axis (I_z) from introductory physics. We notice however, that this is only one component of $\vec{L}$, which also has x- and y-components, $L_x = (-mxz)\omega$ and $L_y = (-myz)\omega$. The quantities $(-mxz)$ and $(-myz)$ have the same units as our familiar moment of inertia (kg·m^2), but do not have the same physical meaning (in fact, they are referred to as *products of inertia*). From this example, we are noticing that there is more to the idea of inertia than the simple scalar quantity from introductory physics, *and* $\vec{L} = I\vec{\omega}$ (where $\vec{L} \parallel \vec{\omega}$) is in general *not* true.

Let's deepen our understanding about the direction of $\vec{L}$ for the rotating IP model of the dancer by using the right-hand rule for cross products. (Recall that to use the right-hand rule for $\vec{L} = \vec{r} \times \vec{p}$, point the fingers of your right hand in the direction of $\vec{r}$, curl them toward the direction of $\vec{p}$ (into the page in Fig. 5.14 (a)), then your thumb points in the direction of $\vec{L}$). We confirm once again that $\vec{L}$ is *not* in the same direction as $\vec{\omega}$. We can also see that the direction of $\vec{L}$ changes as the IP rotates. To see this, imagine the point mass has made one half of a revolution (Fig. 5.14 (b)). Now $\vec{p}$ is coming out of the page (positive x-direction), so $\vec{r} \times \vec{p}$ results in the rotational momentum vector shown. As the IP rotates with a constant rotational velocity around the vertical, the rotational momentum vector changes direction, also rotating around $\hat{z}$. Since $\Sigma\vec{\tau} = \dot{\vec{L}}$, this means $\Sigma\vec{\tau} \neq 0$. Since the change in $\vec{L}$ is in the counterclockwise (CCW) direction, there must be a net CCW torque on the dancer.

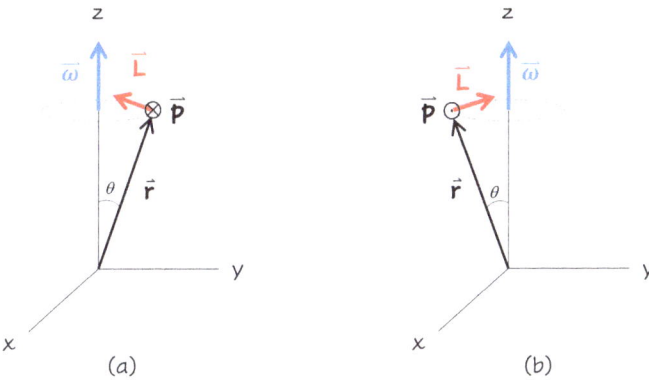

Figure 5.14: The direction of the rotational momentum vector $\vec{L}$ for the inverted pendulum can be found with the right-hand rule for the vector cross product $\vec{L} = \vec{r} \times \vec{p}$. In (a) the point mass is moving into the page and in (b) the IP is moving out of the page. As the IP rotates, $\vec{L}$ also rotates around the z-axis.

In the previous analysis, we treated the dancer as a single point particle at the CM. To extend this analysis and model the dancer as a rigid body consisting of many point masses, we find the total moments and products of inertia by summing the contributions from each individual point mass. For rotation around a single fixed axis ($\vec{\omega} = \omega\hat{z}$), the rotational momentum is:

$$\vec{L} = I_{xz}\omega\hat{x} + I_{yz}\omega\hat{y} + I_{zz}\omega\hat{z}, \quad \text{where}$$
$$I_{xz} = -\Sigma m_i x_i z_i,$$
$$I_{yz} = -\Sigma m_i y_i z_i, \quad \text{and}$$
$$I_{zz} = \Sigma m_i (x_i^2 + y_i^2)$$

where I_{xz} and I_{yz} are the products of inertia and I_{zz} is the moment of inertia about the z-axis. Of course, the axis of rotation for a dancer, or any rigid body, is not in general constrained to a single, fixed axis, and may change with time. For the general case that $\vec{\omega} = \omega_x \hat{x} + \omega_y \hat{y} + \omega_z \hat{z}$, the vector cross product $\vec{L} = \Sigma \vec{r}_i \times \vec{p}_i$ results in the following three components of rotational momentum,

$$L_x = I_{xx}\omega_x + I_{xy}\omega_y + I_{xz}\omega_z$$
$$L_y = I_{yx}\omega_x + I_{yy}\omega_y + I_{yz}\omega_z$$
$$L_z = I_{zx}\omega_x + I_{zy}\omega_y + I_{zz}\omega_z$$

which can be written in matrix form:

$$\vec{L} = \mathbf{I}\vec{\omega}$$

$$\begin{pmatrix} L_x \\ L_y \\ L_z \end{pmatrix} = \begin{pmatrix} I_{xx} & I_{xy} & I_{xz} \\ I_{yx} & I_{yy} & I_{yz} \\ I_{zx} & I_{zy} & I_{zz} \end{pmatrix} \begin{pmatrix} \omega_x \\ \omega_y \\ \omega_z \end{pmatrix} \tag{5.16}$$

The 3×3 matrix,

$$\mathbf{I} = \begin{pmatrix} I_{xx} & I_{xy} & I_{xz} \\ I_{yx} & I_{yy} & I_{yz} \\ I_{zx} & I_{zy} & I_{zz} \end{pmatrix}$$

is the *inertia tensor*. The diagonal elements (I_{xx}, I_{yy}, and I_{zz}) are the *moments of inertia*, with I_{xx} and I_{yy} taking a similar form as I_{zz} found earlier:

$$I_{xx} = \Sigma m_i(y_i^2 + z_i^2)$$
$$I_{yy} = \Sigma m_i(x_i^2 + z_i^2) \tag{5.17}$$
$$I_{zz} = \Sigma m_i(x_i^2 + y_i^2)$$

The off-diagonal elements are the *products of inertia* with

$$I_{xy} = I_{yx} = -\Sigma m_i x_i y_i$$
$$I_{xz} = I_{zx} = -\Sigma m_i x_i z_i \tag{5.18}$$
$$I_{yz} = I_{zy} = -\Sigma m_i y_i z_i$$

Since $I_{xy} = I_{yx}$ and the same relation holds true for the other products of inertia, the inertia tensor always has the property that it is a symmetric matrix. This means that if we reflect the matrix across it's center diagonal (i. e., take the transpose $\mathbf{I}^T$ or exchange rows and columns (row 1 becomes column 1, row 2 becomes column 2, etc.)), the inertia tensor remains unchanged.

$$\mathbf{I} = \mathbf{I}^T$$

5.8.1 Segmental analysis and whole body inertia tensor

To describe the rotational dynamics of a dancer performing whole-body rotations requires determining the inertia tensor for the dancer's body configuration in the turn. We might envision applying eq. (5.17) and eq. (5.18) (or the equivalent for a continuous mass distribution) to find the dancer's whole-body moments and products of inertia. However, we encounter the same types of challenges that we did in Chapter 3 when determining a dancer's whole body center of mass. Since the entire human body is a complex shape, and one that changes with time with relative segmental motion, as discussed in Chapter 3 we would be better served by applying a segmental analysis. We could use eq. (5.17) and eq. (5.18) to find the inertia tensor of *individual* rigid segments of the body (e. g., head, thigh, etc.), making assumptions about their approximate geometric shapes. (An example of this method, treating the thigh as a frustum of a cone, demonstrates this process at the end of this chapter.) However, it is often most practical to use experimental data on segment inertial properties that are reported as averages over populations.

De Leva [20] reports average *radii of gyration* for individual body segments (Tab. 5.1), which can be used to determine the principal moments of inertia relative to the segment's CM using

$$I_j = mr_j^2 \tag{5.19}$$

where m is the segment mass and r_j the radius of gyration, with $j = x'', y'', z''$ being the axes running through the segment CM. When the person stands in anatomical position, the x''-axis runs in the medial-lateral direction (so rotation around x'' would be sagittal plane motion), y'' runs in the anterior-posterior direction (rotations around y'' lead to frontal plane motion), and z'' runs in the longitudinal direction (rotations around z'' lead to transverse plane motion), (see Fig. 2.1). These axes are approximate axes of symmetry of the segments, which is why there are no products of inertia listed in Tab. 5.1.

Example. Use the data in Tab. 5.1 and Tab. 3.1 to find the moment of inertia of a dancer's thigh segment rotated around its longitudinal axis. The dancer's total body mass is 55 kg and her thigh length (from hip joint center to knee joint center) was measured to be 37 cm.

Solution. From Tab. 3.1, we find the mass of the thigh segment, $m = 0.1478M = 8.13$ kg. From Tab. 5.1 and the thigh length, we compute the longitudinal radius of gyration in meters, $r_z'' = 0.162l_{th} = 0.162(0.37 \text{ m}) = 0.060$ m. Finally, the longitudinal moment of inertia of the thigh segment is computed from eq. (5.19), $I_{th,z''} = (8.13 \text{ kg})(0.060 \text{ m})^2 = 0.029$ kg·m^2.

Table 5.1: Body segment radii of gyration (as reported in [20]). Segment endpoints are the same as given in Tab. 3.1. Radii of gyration are reported as a percentage of respective segment length.

Segment	r_x'', F (%)	r_x'', M (%)	r_y'', F (%)	r_y'', M (%)	r_z'', F (%)	r_z'', M (%)
Head	33.0	36.2	35.9	37.6	31.8	31.2
Trunk	35.7	37.2	33.9	34.7	17.1	19.1
Upper Arm	27.8	28.5	26.0	26.9	14.8	15.8
Forearm	26.1	27.6	25.7	26.5	9.4	12.1
Hand	53.1	62.8	45.4	51.3	33.5	40.1
Thigh	36.9	32.9	36.4	32.9	16.2	14.9
Shank	27.1	25.5	26.7	24.9	9.3	10.3
Foot	29.9	25.7	27.9	24.5	13.9	12.4

5.8.2 Modeling a dancer in pirouette position

We can use the data in Tab. 5.1 and from Chapter 3 to determine the whole-body inertia tensor of a dancer in any body configuration. If we know the elements of the inertia tensor for each individual segment (I_{xx}, I_{xy}, I_{xz}, etc., denoted I_{ij}), the elements of the whole-body inertia tensor are found by summing together the contributions from all of the segments ($I_{ij,\text{total}} = I_{ij,\text{head}} + I_{ij,\text{trunk}} + I_{ij,\text{GLfoot}} + I_{ij,\text{SLfoot}} + \cdots$). We have already discussed how to use Tab. 5.1 to find the principal moments of inertia of each segment around the individual segment's CM ($I_{x''}$, $I_{y''}$, and $I_{z''}$). We must instead find the moments and products of inertia relative to a fixed global coordinate system (GCS). The GCS may be both rotated and translated relative to the LCS of the segment. This process will be demonstrated to find the inertia tensor of a dancer in pirouette position. Note that while arduous to do by hand, this process can be automated by use of a computer and can be done using experimental motion capture data (such that total body inertia can be computed dynamically throughout a movement).

Our hypothetical model of a dancer in pirouette position consists of 14 body segments (head, trunk, upper arms, forearms, hands, thighs, shanks, feet) with the origin of the fixed GCS at the ground, labeled (xyz) in Fig. 5.15. The LCS of each body segment has its origin at the segment's CM and is labeled as ($x''y''z''$). The following assumptions were made in creating the model:

1. The CMs of the head and trunk are in line vertically over the GCS origin and the segments are oriented such that $\hat{x} \parallel \hat{x}''$, $\hat{y} \parallel \hat{y}''$, and $\hat{z} = \hat{z}''$.
2. The dancer utilizes 90° hip external rotation (turnout), meaning that for the supporting leg, $\hat{x} \parallel \hat{y}''$, $\hat{y} \parallel \hat{x}''$ and $\hat{z} \parallel \hat{z}''$.
3. The gesture leg (GL) thigh is raised such that its longitudinal axis makes a 60° angle with the vertical.
4. The longitudinal axes of the GL thigh and shank form a 60° angle.
5. The longitudinal axes of the GL shank and foot are aligned.

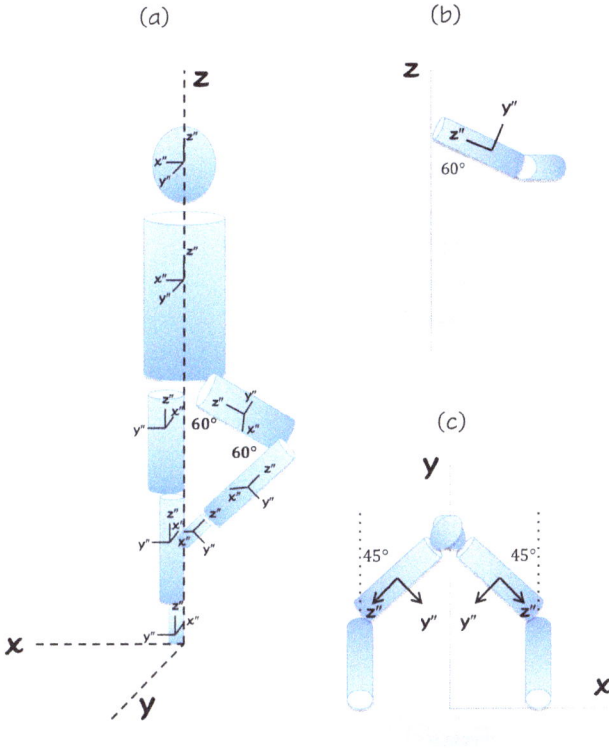

Figure 5.15: A model of a dancer in pirouette position consisting of 14 body segments. (a) Model showing placement and orientation of the head, trunk, supporting leg (thigh, shank, and foot) and gesture leg (thigh, shank, and foot). For simplicity, the arms are not shown here. (b) The placement of the upper arm segment from a side or sagittal plane view (in the *yz*-plane) and (c) the arms from a top down view (in the *xy*-plane).

6. The upper arms are raised (rotated around $\hat{x}''$) such that their longitudinal axes make a 60° angle with the vertical.
7. The forearms are in a plane parallel with the *xy*-plane and are rotated around $\hat{x}''$ by 45° toward the center of the body.

Let's assume the dancer is a male with total body mass $M = 82\,\text{kg}$ and height $H = 1.75\,\text{m}$, and whose body segment lengths were measured. Let's also focus on two example body segments to demonstrate the process for finding the inertia tensor: the supporting leg shank and gesture leg thigh. For each segment we know the inertia in the LCS ($x''y''z''$). For the SL shank, $m_{sh} = 0.0433M = 3.55\,\text{kg}$ and $l_{sh} = 0.436\,\text{m}$, so the inertia tensor in the LCS of the shank is

$$
\mathbf{I}''_{sh} = \begin{pmatrix} m_{sh}r^2_{x''} & 0 & 0 \\ 0 & m_{sh}r^2_{y''} & 0 \\ 0 & 0 & m_{sh}r^2_{z''} \end{pmatrix} = \begin{pmatrix} 0.0439 & 0 & 0 \\ 0 & 0.0418 & 0 \\ 0 & 0 & 0.0072 \end{pmatrix} \text{kg·m}^2
$$

Then, we imagine rotating $(x''y''z'')$ to an intermediate coordinate system $(x'y'z')$ so that $(x'y'z')$ is aligned parallel to the GCS (xyz). Since y'' is parallel to x and x'' is parallel to y (Fig. 5.15), the inertia tensor in our intermediate coordinate system is found simply by interchanging I_{xx} and I_{yy}.

$$\mathbf{I}'_{sh} = \begin{pmatrix} 0.0418 & 0 & 0 \\ 0 & 0.0439 & 0 \\ 0 & 0 & 0.0072 \end{pmatrix} \text{kg·m}^2$$

Finally, we apply the parallel axis theorem noting the displacement of (x, y, z) from (x', y', z'), $\vec{a} = a_x \hat{x} + a_y \hat{y} + a_z \hat{z}$, for this dancer's SL shank is $\vec{a}_{sh} = (-0.054m)\hat{x} + (-0.498m)\hat{z}$. The x-component of displacement is the width of the shank and the z-component of displacement is the longitudinal length of the foot plus the distance from the ankle joint center to the shank's CM. (The shank CM has zero y-displacement.) The general form of the parallel axis theorem that applies to both moments and products of inertia is

$$I_{ij} = I'_{ij} + m(\|\vec{a}\|^2 \delta_{ij} - a_i a_j) \tag{5.20}$$

where we have used compact notation with the Kronecker delta ($\delta_{ij} = 1$ if $i = j$ and $\delta_{ij} = 0$ if $i \neq j$). For example the moment of inertia $I_{xx} = I'_{xx} + m(a_x^2 + a_y^2 + a_z^2 - a_x a_x) = I'_{xx} + m(a_y^2 + a_z^2)$ and the product of inertia $I_{xy} = I'_{xy} - m a_x a_y$ (and similarly for the other moments and products of inertia). From this we obtain the inertia tensor of the SL shank referenced to the GCS axes.

$$\mathbf{I}_{sh} = \begin{pmatrix} 0.9222 & 0 & -0.0955 \\ 0 & 0.9347 & 0 \\ -0.0955 & 0 & 0.0176 \end{pmatrix} \text{kg·m}^2$$

The non-zero x-displacement of the SL shank leads to a small but not zero I_{xz} product of inertia.

Now we move on to the GL thigh (assume $m_{th} = 0.1416M = 11.61$ kg, $l_{th} = 0.4244$ m). The thigh's inertia relative to the LCS origin is

$$\mathbf{I}''_{th} = \begin{pmatrix} m_{th} r_{x''}^2 & 0 & 0 \\ 0 & m_{th} r_{y''}^2 & 0 \\ 0 & 0 & m_{th} r_{z''}^2 \end{pmatrix} = \begin{pmatrix} 0.2263 & 0 & 0 \\ 0 & 0.2263 & 0 \\ 0 & 0 & 0.0464 \end{pmatrix} \text{kg·m}^2$$

To rotate the $(x''y''z'')$ axes of the GL thigh to the intermediate coordinate system $(x'y'z')$ that is parallel with (xyz), we can imagine the dancer taking the passé leg back to anatomical position. To do so, he could first rotate the thigh $-60°$ around the x''-axis (using hip extension to go from 60 degrees to zero degrees of hip flexion), then around the z''-axis by $-90°$ (going from turnout position to parallel with 90° of hip internal rotation). These two rotations in 3D are accomplished with the product of two rotation matrices:

$$\mathbf{S} = \mathbf{R}_{z''}\mathbf{R}_{x''} = \begin{pmatrix} \cos(-90°) & -\sin(-90°) & 0 \\ \sin(-90°) & \cos(-90°) & 0 \\ 0 & 0 & 1 \end{pmatrix} \begin{pmatrix} 1 & 0 & 0 \\ 0 & \cos(-60°) & -\sin(-60°) \\ 0 & \sin(-60°) & \cos(-60°) \end{pmatrix}$$

$$= \begin{pmatrix} 0 & 1 & 0 \\ -1 & 0 & 0 \\ 0 & 0 & 1 \end{pmatrix} \begin{pmatrix} 1 & 0 & 0 \\ 0 & \frac{1}{2} & \frac{\sqrt{3}}{2} \\ 0 & -\frac{\sqrt{3}}{2} & \frac{1}{2} \end{pmatrix} = \begin{pmatrix} 0 & \frac{1}{2} & \frac{\sqrt{3}}{2} \\ -1 & 0 & 0 \\ 0 & -\frac{\sqrt{3}}{2} & \frac{1}{2} \end{pmatrix}$$

The coordinate transformation from $\mathbf{I}''$ to $\mathbf{I}'$ is accomplished through the similarity transformation

$$\mathbf{I}' = \mathbf{S}\mathbf{I}''\mathbf{S}^T \tag{5.21}$$

where $\mathbf{S}^T$ is the transpose of matrix $\mathbf{S}$. For the thigh segment,

$$\mathbf{I}' = \begin{pmatrix} 0 & \frac{1}{2} & \frac{\sqrt{3}}{2} \\ -1 & 0 & 0 \\ 0 & -\frac{\sqrt{3}}{2} & \frac{1}{2} \end{pmatrix} \begin{pmatrix} I_{x''} & 0 & 0 \\ 0 & I_{y''} & 0 \\ 0 & 0 & I_{z''} \end{pmatrix} \begin{pmatrix} 0 & -1 & 0 \\ \frac{1}{2} & 0 & -\frac{\sqrt{3}}{2} \\ \frac{\sqrt{3}}{2} & 0 & \frac{1}{2} \end{pmatrix}$$

$$= \begin{pmatrix} \frac{1}{4}I_{y''} + \frac{3}{4}I_{z''} & 0 & -\frac{\sqrt{3}}{4}I_{y''} + \frac{\sqrt{3}}{4}I_{z''} \\ 0 & I_{x''} & 0 \\ -\frac{\sqrt{3}}{4}I_{y''} + \frac{\sqrt{3}}{4}I_{z''} & 0 & \frac{3}{4}I_{y''} + \frac{1}{4}I_{z''} \end{pmatrix}$$

Numerically for the GL thigh the inertia in the primed coordinate system is

$$\mathbf{I}'_{th} = \begin{pmatrix} 0.0914 & 0 & -0.0779 \\ 0 & 0.2263 & 0 \\ -0.0779 & 0 & 0.1813 \end{pmatrix} \text{kg·m}^2$$

Again, we apply the parallel axis theorem to go from $\mathbf{I}'_{th}$ to $\mathbf{I}_{th}$. We can find the GL thigh's CM displacement in the x-direction with trigonometry. Due to the angle that the GL thigh makes with the vertical, $a_{th,x} = (0.4095l_{th})\sin 60° = 0.151$ m. (Note that $0.4095l_{th}$ is the distance of the thigh CM from the hip joint center along its longitudinal axis.) The thigh is displaced vertically by the lengths of the foot, shank and thigh, minus the vertical distance of the thigh CM from the HJC. So $a_{th,z} = -(l_f + l_{sh} + l_{th} - ((0.4095l_{th})\cos 60°)) = -1.027$ m. Recall that $\vec{a}$ is the displacement as we move from the primed to unprimed coordinate system, so we are moving in the $+x$- and $-z$-directions to go from the GL thigh CM to the ground (Fig. 5.15). Using this value of $\vec{a}_{th} = (0.151m)\hat{x} + (-1.027m)\hat{z}$ and the parallel axis theorem given in eq. (5.20), we have

$$\mathbf{I}_{th} = \begin{pmatrix} 12.34 & 0 & 1.723 \\ 0 & 12.74 & 0 \\ 1.723 & 0 & 0.446 \end{pmatrix} \text{kg·m}^2$$

Given that the GL thigh is further from all of the global axes, it makes sense that its moments and products of inertia are larger than that which we found for the SL shank. To find the total inertia of the dancer, we repeat this process for all of the individual segments, then sum their individual inertias ($I_{ij,total} = I_{ij,head} + I_{ij,trunk} + I_{ij,GLfoot} + I_{ij,SLfoot} + \cdots$). Doing this we find that the total inertia of the male dancer ($M = 82\,\text{kg}$, $H = 1.75\,\text{m}$) in passé pirouette position is

$$\mathbf{I} = \begin{pmatrix} 139.8 & 0 & 1.09 \\ 0 & 139.8 & -3.11 \\ 1.09 & -3.11 & 1.87 \end{pmatrix} \text{kg·m}^2$$

The inertia tensor for the dancer in pirouette position has small, nonzero products of inertia, meaning if the dancer rotates around the vertical (global z) their rotational momentum vector and rotational velocity vectors will not quite be parallel. Like the rotating inverted pendulum discussed previously, there will be a net torque changing the direction of $\vec{L}$ as the dancer rotates, leading to a slight wobble. This means that the dancer would not be dynamically balanced. No matter the axis of rotation, however, now that we know the inertia tensor for the dancer, we could apply rigid body dynamics ($\Sigma\vec{\tau} = \frac{d}{dt}(\mathbf{I}\vec{\omega})$) for the general 3D rotation of a dancer in a pirouette. We could also repeat this process for any body configuration for different styles of turn. It's worth noting that the equations of motion can become very complicated when the chosen coordinate system is not that of the principal axes. The meaning of principal axes and how to solve for them are discussed in the next section.

5.9 Principal moments of inertia, principal axes, and dynamic balance during rotations

We have seen that in general the rotational momentum and velocity vectors of a rigid rotating body may not be in the same direction. In a case in which $\vec{L}$ *is* parallel to $\vec{\omega}$, $\vec{L} = \mathbf{I}\vec{\omega}$ becomes

$$\vec{L} = \lambda\vec{\omega} \tag{5.22}$$

where λ is a scalar. An axis of rotation for which eq. (5.22) holds true is called a *principal axis*, and the quantity λ is a *principal moment of inertia*. Before delving into the mathematical details, let's once again develop some conceptual understanding by revisiting our rotating simple IP (Fig. 5.14). Notice if instead of a single point mass, our rotating object consists of two identical IPs reflected across the z-axis, we effectively cancel the *x*- and *y*-components of $\vec{L}$ as the two IPs rotate together. Then $\vec{L} \parallel \vec{\omega}$ as the object rotates (Fig. 5.16). Now the *z*-axis is a principal axis for this two mass IP system.

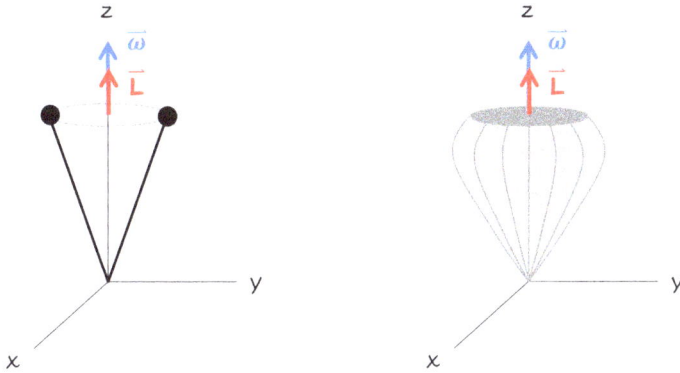

Figure 5.16: When a rotationally symmetric object is rotated around its axis of symmetry, $\vec{L} \parallel \vec{\omega}$, meaning the symmetry axis is a principal axis of rotation. Three orthogonal principal axes of rotation exist for all rigid bodies, whether or not they exhibit rotational symmetry.

Let's compute the products of inertia of the two mass IP system. With the masses in the configuration on the left of Fig. 5.16, $x_1 = x_2 = 0$, $y = y_1 = -y_2$, and $z = z_1 = z_2$. From eq. (5.17), the moments of inertia are $I_{xx} = 2m\rho_x^2 = 2m(y^2 + z^2)$, $I_{yy} = 2m\rho_y^2 = 2mz^2$, and $I_{zz} = 2m\rho_z^2 = 2my^2$. Eq. (5.18) for the products of inertia give us

$$I_{xy} = -(m(0)y + m(0)(-y)) = 0$$
$$I_{xz} = -(m(0)z + m(0)z) = 0$$
$$I_{yz} = -(myz + m(-y)z) = 0$$

With all of the products of inertia equal to zero, the inertia tensor is a diagonal matrix:

$$\mathbf{I} = \begin{pmatrix} \lambda_1 & 0 & 0 \\ 0 & \lambda_2 & 0 \\ 0 & 0 & \lambda_3 \end{pmatrix}$$

Now for rotation around any axis ($\omega = \omega_x \hat{x} + \omega_y \hat{y} + \omega_z \hat{z}$), when the inertia tensor is diagonalized, $\vec{L} = \mathbf{I}\vec{\omega}$ results in $\vec{L} = \lambda_1 \omega_x \hat{x} + \lambda_2 \omega_y \hat{y} + \lambda_3 \omega_z \hat{z}$. From this result, we also see that if $\vec{\omega} = \omega \hat{x}$, $\vec{L} = \lambda_1 \omega \hat{x}$, if $\vec{\omega} = \omega \hat{y}$, $\vec{L} = \lambda_2 \omega \hat{y}$, and if $\vec{\omega} = \omega \hat{z}$, $\vec{L} = \lambda_3 \omega \hat{z}$. This means that *x*, *y*, and *z* are principal axes.

Another consequence of rotation around a principal axis is that the general form of Newton's second law for rotations (eq. (5.5)) reduces to the simplified form from introductory physics.

$$\Sigma\vec{\tau} = \frac{d\vec{L}}{dt} = \frac{d}{dt}(\lambda\vec{\omega}) = \lambda\frac{d\vec{\omega}}{dt}, \quad \text{so}$$

$$\Sigma\vec{\tau} = \lambda\vec{\alpha} \quad \text{for rotation around a principal axis.}$$

When an object is rotationally symmetric (e. g., the two mass IP system, a cone, a sphere, etc.), principal axes can be straightforward to find, since one principal axis will be the symmetry axis. The other two principal axes will be orthogonal to the symmetry axis (as well as eachother). For the shape like a top (on the right in Fig. 5.16), x- and y- can be rotatated any amount around z- and remain principal axes due to the top's symmetry. This means any orthogonal axes in the xy-plane would serve as principal axes in the case of the symmetric top.

Even if an object is not rotationally symmetric (e. g., a dancer in pirouette position), principal axes exist for *all* rigid bodies. And even if the object has a strange shape, when rotated around a principal axis, $\vec{L} \parallel \vec{\omega}$, and the object will be dynamically balanced. To find the principal axes (and principal moments) becomes an eigenvalue problem. For principal axes,

$$\mathbf{I}\vec{\omega} = \lambda\mathbf{1}\vec{\omega}$$

$$(\mathbf{I} - \lambda\mathbf{1})\vec{\omega} = 0$$

where $\mathbf{1}$ is the identity matrix. Unless $\vec{\omega} = 0$, which is uninteresting,

$$\det(\mathbf{I} - \lambda\mathbf{1}) = 0$$

Since $\mathbf{I}$ is a 3×3 tensor, we will have three eigenvalues (λ_1, λ_2, and λ_3) for the three principal moments of inertia, and three eigenvectors ($\hat{e}_1$, $\hat{e}_2$, and $\hat{e}_3$) corresponding to those eigenvalues, which are the principal axes.

We can now find the principal moments of inertia and principal axes of rotation of the dancer in pirouette position from Section 5.8.2. (The eigenvalues and eigenvectors can be found by hand or with a program, such as Mathematica.)

$$\lambda_1 = 139.9 \text{ kg·m}^2, \quad \lambda_2 = 139.8 \text{ kg·m}^2, \quad \lambda_3 = 1.79 \text{ kg·m}^2$$

$$\hat{e}_1 = \begin{pmatrix} 0.33 \\ -0.94 \\ -0.02 \end{pmatrix}, \quad \hat{e}_2 = \begin{pmatrix} 0.94 \\ 0.33 \\ 0.00 \end{pmatrix}, \quad \hat{e}_3 = \begin{pmatrix} -0.01 \\ 0.02 \\ 1.00 \end{pmatrix}$$

Here we make several observations. The third principal axis, $\hat{e}_3$ is nearly in the $\hat{z}$-direction, which we will refer to as the spin principal axis. We also notice that $\hat{e}_1$ is not quite in the $-\hat{y}$-direction, and $\hat{e}_2$ is not quite in the $\hat{x}$-direction. These are the two principal topple axes, with rotation of the body around $\hat{e}_1$ corresponding to topple in (approximately) the medial-lateral direction, and $\hat{e}_2$ corresponding to topple in (approximately) the anterior-posterior direction. The good news for the dancer attempting triple turn pirouettes (or any dancer) is that the vertical axis is nearly already a principal axis for pirouette position.

The moment of inertia corresponding to the spin principal axis, $I_3 = 1.79 \, \text{kg·m}^2$, is much less than the other two principal moments of inertia, which makes sense physically. In pirouette position, the body's mass is distributed closer to the spin axis than the topple axes. We also notice that $I_1 \approx I_2$, meaning that based on distribution of mass around these axes, the propensity to topple in either direction is roughly equal. If a dancer notices that they consistently topple one direction in a standard ballet pirouette, it is not because of the shape or distribution of mass. It is most likely due to the manner in which the dancer initiates the pirouette during pushoff.

5.10 Body segment rotations around joint axes

Whether or not the entire body rotates, dancing consists of many rigid body rotations as individual body segments rotate around joints. Often in biomechanics, we are interested in measuring the angles between two segments connected at a joint and not just the rotation of one of the segments. *Joint angles* provide information about the orientation of one segment relative to its adjoining segment. For example, the knee angle is the 3D orientation of the shank relative to the thigh. As learned in Chapter 2, different joint types allow differing ranges of motion in various directions. The knee acts like a hinge joint with its primary range of motion in the flexion/extension direction (rotating around the joint's x''-axis). However, there is still some range of motion in the ab/adduction (y'') and internal/external rotation (z'') directions. For a ball and socket joint like the hip, the relative rotation of the thigh and pelvis segments have large range of motion around all three joint axes of rotation.

Because joint angles are the orientation of one segment relative to another, in reality joint angle information is contained in a 3D rotation matrix. This is much like the rotational transformation matrices utilized in Section 5.8.2 to go from the rotated $(x''y''z'')$ frame to the unrotated $(x'y'z')$ frame of the gesture leg thigh. For the knee joint angle, if we perform a rotational transformation from the shank's local coordinate system to align it with the thigh's local coordinate system, the transformation matrix gives us joint angle information.

One very important feature of rotational transformations is that the order in which they are performed matters. It's not enough to say the thigh segment was rotated around its own x''-axis and its own z''-axis; we must say which rotation occurred first. Imagine beginning in anatomical position. Performing 90° of hip external rotation would rotate the thigh around z'' (and place the legs in ballet first position with turnout). Then a rotation around the thigh's x''-axis by 90° lifts the leg into à la seconde (leg held out to the side). Now let's start in anatomical position and this time rotate the thigh around x'' and then z''. From AP, rotation of the thigh around x'' by 90° produces hip flexion, so the leg is held in front of the body. Then z'' rotation of 90°, moves the leg from parallel to turned out position while still being held in front of the body. The different ordering of rotations made it such that the leg was held in completely different final positions (devant or à la seconde).

To report out three joint angles instead of a 3×3 transformation matrix, one must specify the angles rotated around the various axes and in what order. A certain sequence of rotations, the Cardan sequence ($x''-y''-z''$), is often used in biomechanics research, and its output angles give the amount of flexion/extension (x''), ab/adduction (y''), longitudinal rotation (z'') at a joint. Computations of these joint angles from experimental data can be automated by researchers. The specific details of performing the coordinate transformations will not be provided here, but it is important for researchers to understand how these calculations are performed, which series of rotations were chosen, and the meaning of the three joint angles that are output. Standards for the process of measuring and reporting joint angles (as well as other biomechanical quantities) have been proposed by the International Society of Biomechanics (ISB), for example [109] and [110].

How the motion of a body segment (or joint angle) changes under the influence of a net torque is influenced by muscle forces acting on a segment. This process is addressed in Chapter 6 on Musculoskeletal Forces. The dynamics of segmental rotation around a joint also requires knowledge about the inertia of the segment(s) of interest around the joint. As we know, we can use the data on radii of gyration in Tab. 5.1. Let's also take this opportunity to compute radii of gyration from a rough geometric model of a segment to confirm that the values given in Tab. 5.1 make sense.

Example. As done in Chapter 3, let's model the thigh segment as a frustum of a cone (Fig. 3.4) with radius of the proximal end R_P, the distal end R_D, the length of the thigh L, and its mass m_{th}. Find the inertia tensor of the thigh segment rotated around the hip joint.

Solution. We will use the coordinate system in Fig. 3.4 with the origin at the proximal endpoint (hip). In Cartesian coordinates, the inertia tensor around the origin is:

$$
I = \begin{pmatrix} I_{xx} & I_{xy} & I_{xz} \\ I_{yx} & I_{yy} & I_{yz} \\ I_{zx} & I_{zy} & I_{zz} \end{pmatrix}
$$

$$
= \int \begin{pmatrix} (y^2 + z^2) & -xy & -xz \\ -xy & (x^2 + z^2) & -yz \\ -xz & -yz & (x^2 + y^2) \end{pmatrix} dm
$$

$$
= \rho \int \begin{pmatrix} (y^2 + z^2) & -xy & -xz \\ -xy & (x^2 + z^2) & -yz \\ -xz & -yz & (x^2 + y^2) \end{pmatrix} dV
$$

Assuming uniform mass distribution, the mass density of the thigh is then:

$$
\rho = \frac{m_{\text{th}}}{V_{\text{th}}} = \frac{3 m_{\text{th}}}{\pi L (R_D^2 + R_D R_P + R_P^2)}
$$

where we have used the total volume of the frustum found in Chapter 3, $V_{th} = \frac{\pi L}{3}(R_D^2 + R_D R_P + R_P^2)$. The radius of the thigh varies linearly as a function of z (where $z = 0$ at the proximal end), with

$$r(z) = R_P + (R_D - R_P)\frac{z}{L}$$

as found in Chapter 3. Using cylindrical coordinates ($x = r\cos\phi$, $y = r\sin\phi$, $z = z$), and $dV = r\,dr\,d\phi\,dz$, the inertia tensor can be rewritten as

$$I = \rho \int_{z=0}^{L}\int_{\phi=0}^{2\pi}\int_{r=0}^{r(z)} \begin{pmatrix} (r\sin\phi)^2 + z^2 & -r^2\sin\phi\cos\phi & -(r\cos\phi)z \\ -r^2\sin\phi\cos\phi & (r\cos\phi)^2 + z^2 & -(r\sin\phi)z \\ -(r\cos\phi)z & -(r\sin\phi)z & r^2 \end{pmatrix} r\,dr\,d\phi\,dz$$

The products of inertia all have $\sin\phi$ or $\cos\phi$ integrated over one full cycle from 0 to 2π, so we find that

$$I_{xy} = I_{yx} = I_{xz} = I_{zx} = I_{yz} = I_{zy} = 0$$

which is our expected result, since the z-axis (longitudinal axis) of the frustum is an axis of rotational symmetry. Substituting ρ and $r(z)$, the moments of inertia of the thigh are

$$I_{xx} = I_{yy} = m_{th}\frac{2L^2(6R_D^2 + 3R_D R_P + R_P^2) + 3(R_D^4 + R_D^3 R_P + R_D^2 R_P^2 + R_D R_P^3 + R_P^4)}{20(R_D^2 + R_D R_P + R_P^2)}$$

$$I_{zz} = m_{th}\frac{3(R_D^4 + R_D^3 R_P + R_D^2 R_P^2 + R_D R_P^3 + R_P^4)}{10(R_D^2 + R_D R_P + R_P^2)}$$

which can now be used to find numerical values given dimensional measurements of the thigh for an individual dancer. For example, if measurements of a female dancer's thigh were $R_D = 0.05$ m, $R_P = 0.075$ m, and $L = 0.35$ m, then $I_{xx} = I_{yy} = m_{th}(0.0339)$ in units of kg·m². From eq. (5.19), this means that $r_{xx}^2 = r_{yy}^2 = 0.0339$ m², or the radii of gyration from the hip joint are $r_{xx} = r_{yy} = \sqrt{0.0339\text{ m}^2} = 0.184$ m. We divide by L to report those as a fraction of the thigh's length, or $\frac{r_{xx}}{L} = \frac{r_{yy}}{L} = 0.526$. The same process is followed to determine r_{zz} and we find that

$$r_{\text{gyration}} = \begin{pmatrix} 52.6\,\% & 0 & 0 \\ 0 & 52.6\,\% & 0 \\ 0 & 0 & 13.0\,\% \end{pmatrix} \quad \text{from proximal end}$$

The sagittal and transverse radii are not very close to de Leva's values for the thigh reported in Table 5.1, which are $r_x'' = 36.9\,\%$, $r_y'' = 36.4\,\%$, and $r_z'' = 16.2\,\%$. This is because de Leva reports the radii of gyration with the origin at the CM of the thigh while we measured from the proximal end. Using parallel axis theorem, we can find the inertia tensor around the CM position, which gives a calculated radii of gyration that are close to those given by de Leva in Tab. 5.1:

$$r_{\text{gyration}} = \begin{pmatrix} 35.6\,\% & 0 & 0 \\ 0 & 35.6\,\% & 0 \\ 0 & 0 & 17.7\,\% \end{pmatrix} \quad \text{from the thigh's CM}$$

6 Musculoskeletal forces

A ballet dancer has noticed that the pain he has been experiencing – and trying his best to ignore – toward the top of the shin, right below the knee has continued to worsen. He observes that the lingering pain is exacerbated when he lands from jumps, and the dancer is beginning to worry that he is developing patellar tendinitis like one of his fellow dancers. So far, the pain has not affected his dancing, but if it worsens, his performance may experience negative consequences. The dancer wonders, "What could be causing this pain?" And, "What can I do to stop the pain and to prevent it from reoccurring?"

Our study of forces thus far has been restricted to forces produced by sources outside of the body while dancing; however, forces produced inside of the body, like those experienced by the patellar ligament of this dancer, are important to consider and may lead to pain or injury. This dancer may have patellar tendinitis, which is inflammation in the ligament connecting the patella (knee cap) to the tibia (shin bone), caused by tiny microtears when the ligament is under repeated stress. When caught early, icing and increased rest between repetitive movements can allow the microtears to heal. However, if the microtears aren't given time to heal, tendonitis can worsen into tendinosis, or tissue (collagen) degeneration. Recovery time can go from weeks for tendinitis to months for tendinosis and require more difficult treatment for the collagen fibers to repair. Thankfully in this scenario, the dancer caught the tendinitis early, worked out a treatment plan with a physical therapist, and talked with his dance teacher about modifying his training load while he recovered. So it was patellar tendinitis causing the pain, but what caused the tendinitis in the first place and how can the dancer prevent the pain from returning?

One possible cause could have been a muscle imbalance, e. g., if one of the quadriceps muscles pulls with too much or too little force and the overall effect causes the muscle-tendon complex to pull on the patella at an unideal angle. The patella moves along a groove in the femur and a muscle imbalance can cause it to improperly track medially or laterally. Poor alignment of the lower limbs can also lead to improper tracking of the patella and a host of other issues at the knee and/or ankle joints. A common source of poor alignment in ballet dancers occurs when they "force" their turnout. Ballet technique requires dancing with turnout or external rotation at the hip joint such that the entire lower limb, including the feet, turn outward. The external rotation should ideally only occur at the hip joint, achieved by using the muscles of the hip's lateral rotator group. Dancers who are unable to achieve optimal turnout from the hips might compensate by forcing the additional turnout at the feet, which are held in place with friction at the floor. Dancing with forced turnout via external rotation at the knee or ankle joints may lead to issues such as weakening the ankle ligaments, foot pronation, and pain in the ankles and/or knees. Forcing turnout at the feet may produce a cascade of negative effects along the lower limbs and poor pelvis alignment or lumbar lordosis. All of these alignment issues can lead to increased joint forces while dancing.

https://doi.org/10.1515/9783110642292-006

Dance teachers can be quite adept at catching improper technique or alignment issues such as those caused by forced turnout. Visually inspecting the technique of dancers during dynamic movements or in large groups can be challenging, however. Some issues are internal and more difficult or impossible to see from the outside, such as muscle activation patterns or coactivation. Noninvasive, direct measurements of muscle and joint forces during dancing would be a great benefit to dancers and dance teachers. Dancers who suffer recurring injuries could determine whether certain muscle forces are too large, or not large enough, while they dance and may lead to increased ligamentous forces or large joint forces. Dance teachers could test whether certain changes to technique or verbal coaching could modify musculoskeletal forces their dancers experience. If correlations between muscle force and injury are discovered, dancers could be screened to prevent injuries before they occur.

However, technology does not currently exist to measure forces inside of the body non-invasively. Force transducers can be implanted into tendon to directly measure force, however the surgical procedure is invasive and studies are primarily limited to non-human animals. Electromyography (EMG) using needle electrodes or less invasive surface electrodes can measure muscle activation during dynamic movements, but these measurements do not scale directly to muscle force.

There is an *indirect* method to quantify musculoskeletal forces, known as inverse dynamics. Knowing the kinematics of body segments and their inertial properties, the net torque on the segment can be determined. To isolate and solve for individual musculoskeletal forces that contribute to the net torque require biomechanical models built with anatomical knowledge, such as muscle attachment points, lines of action of muscle forces, etc. The following sections will lead us through this inverse dynamics approach and discuss the usefulness, challenges, and limitations of this technique to quantify forces inside of the body.

6.1 Inverse dynamics

In previous chapters, we applied Newton's second law to a system of interest with the goal of predicting motion from known applied forces or torques. When analyzing a dancer's CM motion during the push-off for a jump in Chapter 3, we experimentally measured the external forces, $\overrightarrow{GRF}$ and $\vec{F}_g$. From net force and known initial conditions ($\vec{v}_0$ and $\vec{x}_0$), kinematic variables with respect to time can be predicted.

$$\text{measure } \Sigma\vec{F}(t) \;\rightarrow\; \frac{\Sigma\vec{F}(t)}{m} = \vec{a}(t) \;\rightarrow\; \int_{t_0}^{t} \vec{a}(t)\mathrm{d}t = \vec{v}(t) - \vec{v}_0 \;\rightarrow\; \int_{t_0}^{t} \vec{v}(t)\mathrm{d}t = \vec{x}(t) - \vec{x}_0$$

This approach can also be taken to determine rotational kinematics from the net torque on a system. The process of using the forces and torques (the "cause") to determine the kinematics (the "effect") is termed *forward dynamics*.

When faced with the problem of wishing to know forces and torques inside of the body that are unable to be measured directly (e. g., tension forces in muscle or tendon, compressive forces on bone, etc.), we can instead take a different approach based on the quantities that we *can* measure. In this case the system we wish to analyze is an individual body segment; therefore, some of the *external forces on the system* are forces produced *inside of the body*. For example, if a dancer lifts an arm to the side with shoulder abduction, one of the external forces acting on the arm system is the deltoid muscle force (a primary shoulder abductor), which attaches to the humerus (upper arm) across the shoulder joint. In a reverse process compared to forward dynamics, if we measure the 3D position, $\vec{x}(t)$, and orientation, $\vec{\theta}(t)$, of the system throughout a movement, we can use those kinematics and the inertial properties of our system to solve for net force and net torque.

$$\text{measure } \vec{x}(t) \rightarrow \frac{d\vec{x}(t)}{dt} = \vec{v}(t) \rightarrow \frac{d\vec{v}(t)}{dt} = \vec{a}(t) \rightarrow m\vec{a}(t) = \Sigma\vec{F}(t) \tag{6.1}$$

$$\text{measure } \vec{\theta}(t) \rightarrow \frac{d\vec{\theta}(t)}{dt} = \vec{\omega}(t) \rightarrow I\vec{\omega} = \vec{L}(t) \rightarrow \frac{d\vec{L}(t)}{dt} = \Sigma\vec{\tau}(t) \tag{6.2}$$

This process of using the kinematics ("effect") to determine the net force and net torque ("cause") is termed *inverse dynamics*. When we apply the inverse dynamics process in eq. (6.1) and eq. (6.2), we still have not determined a specific musculoskeletal force, but the total force and total torque on the segment.

Example. Let's take a closer look at the example of a dancer lifting an arm to the side to demonstrate the inverse dynamics approach. Kinematic data of body segments can be experimentally measured with a motion capture system; however, in this example, we will use a purely theoretical approach to devise an expression for angular position as a function of time for the arm, as an exercise in choosing reasonable assumptions. Let's assume the dancer begins with the arm held at rest ($\omega_0 = 0$ rad/s) vertically downward with zero degrees of shoulder abduction ($\theta_0 = 0$ rad) and then moves the arm to a final position with 90° of shoulder abduction ($\theta_f = \frac{\pi}{2}$ rad) where the arm comes to rest again ($\omega_f = 0$ rad/s). We will also assume the duration of this movement is two seconds, a reasonable pace for a dancer. It would also be reasonable to assume that if the arm is moved smoothly that the maximum velocity of the arm occurs at the midpoint and a plot of θ vs. t would look something like Fig. 6.1. You may notice that this resembles half of a cycle of a shifted cosine function.

(a) Assume that $\theta(t)$ takes the functional form of $\theta(t) = A\cos(2\pi ft) + B$. Use the information given above to determine the numerical values of A, B, and f and therefore θ as a function of time for the arm lift.

(b) Apply the inverse dynamics approach to determine the net torque on the dancer's arm as a function of time. Assume that the dancer's arm is a long cylinder with total length $l = 62$ cm, rotated around one end and the dancer's total body mass is 60 kg.

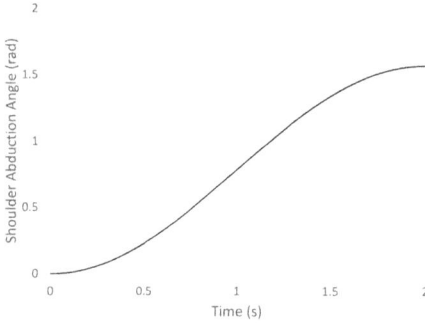

Figure 6.1: Shoulder abduction angle as a function of time during port de bras (arm lift). Note that the angle is given in radians and not degrees, and $\theta_{max} = 1.57\,\text{rad} = 90°$.

Solution. (a) Since one-half of a cycle is achieved in two seconds, one period, T, or time for one full cycle is four seconds. The frequency is the number of cycles per second, so

$$f = \frac{1}{T} = \frac{1}{4}\,\text{Hz}$$

Therefore, $\theta(t) = A\cos(2\pi(\frac{1}{4})t) + B = A\cos(\frac{\pi}{2}t) + B$. We then apply the initial and final conditions, $\theta_0 = 0\,\text{rad}$ and $\theta_f = \frac{\pi}{2}\,\text{rad}$, to determine the values of the constants A and B.

$$\theta(t=0) = 0 = A\cos(0) + B \rightarrow A + B = 0$$

$$\theta(t=2) = \frac{\pi}{2} = A\cos(\pi) + B \rightarrow -A + B = \frac{\pi}{2}$$

Combining the two expressions results in $A = -\frac{\pi}{4}$ and $B = \frac{\pi}{4}$. Therefore, the shoulder abduction angle as a function of time during the arm lift is

$$\theta(t) = -\frac{\pi}{4}\cos\left(\frac{\pi}{2}t\right) + \frac{\pi}{4}.$$

(b) Now that we have an expression for $\theta(t)$ we can apply the inverse dynamics approach in eq. (6.2). We first differentiate $\theta(t)$ with respect to time to find $\omega(t)$:

$$\omega(t) = \frac{d\theta}{dt} = \frac{d}{dt}\left(-\frac{\pi}{4}\cos\left(\frac{\pi}{2}t\right) + \frac{\pi}{4}\right) = \frac{1}{8}\pi^2\sin\left(\frac{\pi}{2}t\right).$$

We find the arm's rotational momentum as a function of time $L(t) = I\omega(t)$ by first determining the moment of inertia of the arm. Modeling the arm as a long cylinder rotated around one end, $I = \frac{1}{3}ml^2$. So,

$$L(t) = I\omega = \frac{1}{3}ml^2\left(\frac{1}{8}\pi^2\sin\left(\frac{\pi}{2}t\right)\right) = \frac{1}{24}\pi^2 ml^2\sin\left(\frac{\pi}{2}t\right).$$

And finally,

$$\sum \tau(t) = \frac{dL(t)}{dt} = \frac{1}{48}\pi^3 ml^2 \cos\left(\frac{\pi}{2}t\right).$$

To find numerical values of net torque, we need values of m and l. We are given that $l = 0.62\,\text{m}$ and we use Table 3.1 to find the mass of one arm (upper arm, forearm, and hand), $m = (0.0255 + 0.0138 + 0.0056)M = 0.0449M$ for a female dancer, where M is the dancer's total body mass (60 kg). Substituting these values, we find

$$\sum \tau(t) = 0.669 \cos\left(\frac{\pi}{2}t\right)$$

The results for the arm's rotational velocity around the shoulder and net torque versus time are displayed in Fig. 6.2. The analysis suggests that maximum positive net torque occurs as the movement is initiated ($t = 0$) and maximum negative torque occurs at the end of the movement, at peak shoulder abduction. In the next example we will compare the results from our rough model to that found using actual experimental data.

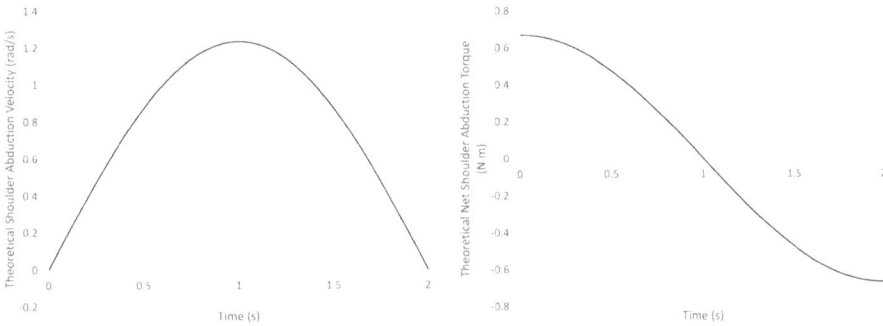

Figure 6.2: Shoulder abduction rotational velocity (left) and net torque on the arm segment (right) as predicted by the theoretical approximation.

Example. A dancer ($M = 60\,\text{kg}$) does an arm lift, and her shoulder abduction angle was measured experimentally throughout the movement (sampling frequency $= 100\,\text{Hz}$). Shoulder abduction angle as a function of time is displayed in Fig. 6.3.

(a) Compare the experimentally obtained data to the theoretical $\theta(t)$ from the previous example.

(b) Use the finite differences method to determine the experimental $\omega(t)$ and compare to the theoretically derived expression.

(c) Find the moment of inertia of the arm segment abducted around the shoulder joint by using de Leva data for segment masses (Tab. 3.1), center of mass positions (Tab. 3.2), and radii of gyration (Tab. 5.1). The lengths of the arm segments (upper arm (UA), forearm (FA), and hand (H)) were measured to be $l_{\text{UA}} = 27.7\,\text{cm}$, $l_{\text{FA}} = 26.5\,\text{cm}$, and $l_{H} = 7.8\,\text{cm}$.

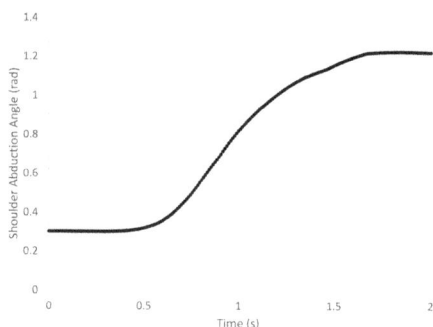

Figure 6.3: Experimentally measured shoulder abduction angle during an arm lift. The data take a similar, but not exact, shape as the theoretical prediction in Fig. 6.1.

(d) Find and plot the experimentally determined $\sum \tau(t)$. Interpret and compare to that theoretically obtained in the previous example.

Solution. (a) The shape of the experimental $\theta(t)$ (Fig. 6.3) looks quite similar to the theoretical approximation in Fig. 6.1. While not quite as smooth, it is still a fairly smooth curve. One noticeable difference is that the experimental arm lift began with a non-zero shoulder abduction angle, with the arm held at rest initially ($\theta_0 = 0.3\,\text{rad} \approx 17°$), and the final experimental shoulder abduction angle was $\theta_f = 1.22\,\text{rad} \approx 70°$ instead of $90°$.

(b) Experimental rotational velocity as a function of time is shown in Fig. 6.4. Like the theoretical prediction, it begins and ends with $\omega = 0$, is positive for all times as the arm is lifted, and its maximum is near the midpoint. The single, smooth peak in the theoretical plot is more sharp experimentally, and the experimental data are overall less smooth.

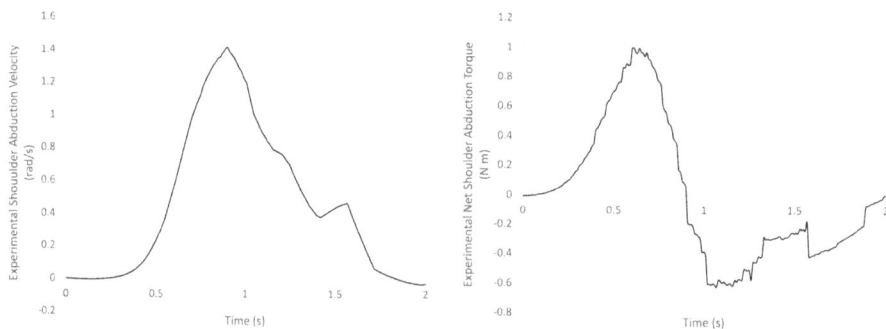

Figure 6.4: Experimentally measured rotational velocity (left) and net torque on the arm (right) during shoulder abduction. Both similarities and differences exist between these data and the predictions made in Fig. 6.2.

(c) To find the total moment of inertia of the arm segment around the shoulder joint during abduction, we first use the de Leva data with total body mass and segment lengths given in the example to compute the segment masses, segment CM positions, and transverse radii of gyration (r_y'', since shoulder abduction occurs in the transverse plane). The results are summarized in Tab. 6.1.

Table 6.1: Dancer's arm segment data summary table. CM position is measured from the segment's proximal endpoint. Transverse r is the radius of gyration from the segment's CM.

Segment	mass (kg)	length (cm)	CM position (cm)	transverse r (cm)
Upper Arm (UA)	1.530	27.7	15.94	7.202
Forearm (FA)	0.828	26.5	12.08	6.811
Hand (H)	0.336	7.8	5.83	3.541

We then compute the transverse moment of inertia of each individual segment around the shoulder joint. Since the radii of gyration (r) are relative to the respective segment's CM position, the parallel axis theorem is applied, $I = mr^2 + md^2$, where d is the displacement of the segment's CM from the shoulder joint. Using the information in Tab. 6.1, $d_{UA} = 15.94$ cm, $d_{FA} = (27.7 + 12.08)$ cm $= 39.78$ cm, and $d_H = (27.7 + 26.5 + 5.83)$ cm $= 60.03$ cm. Using the respective masses, radii of gyration, and the parallel axis theorem, we find that

$$I_{UA} = 0.0468 \text{ kg·m}^2; \quad I_{FA} = 0.1349 \text{ kg·m}^2; \quad I_H = 0.1215 \text{ kg·m}^2$$
$$I = I_{UA} + I_{FA} + I_H = 0.303 \text{ kg·m}^2$$

Comparing this value to our crude model of the arm as a uniform, long thin cylinder rotated around one end ($I = \frac{1}{3}ml^2 = \frac{1}{3}(2.694 \text{ kg})(0.62 \text{ m})^2 = 0.345 \text{ kg·m}^2$), we find that the values are somewhat close, with the long cylinder model about 14 % larger than that found using the de Leva data. This makes sense, given that the long cylinder approximates a uniform mass distribution, with as much mass located further from the axis of rotation (distally) as there is closer to the axis of rotation (proximally). In reality, more mass is concentrated closer to the axis of rotation (the shape of the arm tapers from the upper arm to the hands), so the true inertia will be lower than that calculated with the long cylinder model.

(d) We now can compute the rotational momentum as a function of time from $\vec{L} = I\vec{\omega}$ and apply the finite difference method once again to determine $\frac{d\vec{L}}{dt}$, which is equal to the net torque on the arm (eq. (6.2)). Although the data were filtered, we notice more difference between the experimental torque (Fig. 6.4) and theoretical torque (Fig. 6.2) than we did in our other comparisons. Overall, both calculations of torque show $\Sigma\vec{\tau} > 0$ in approximately the first half of the motion, then $\Sigma\vec{\tau} < 0$ in the second half. The peak positive acceleration theoretically occurred as soon as the motion commenced ($t = 0$), whereas in reality, it took some time for the peak torque to develop. In the theoretical example, going from zero to maximum torque at $t = 0$ and then from maximum negative torque to zero again at the end of the motion ($t = 2\,\text{s}$) is a discontinuity that does not occur in real human motion.

6.2 Net musculoskeletal force and torque

The net force and net torque on a segment are directly proportional to the segment's linear and rotational acceleration, respectively; however, these quantities do not yet provide us with biomechanically meaningful information such as the magnitudes of the *musculoskeletal* forces and torques responsible for producing the motion. To get a step closer to more meaningful biomechanical quantities, we can separate the net force and net torque in eq. (6.1) and eq. (6.2) into two terms. The net force, $\sum \vec{F}(t) = \sum \vec{F}(t)_{\text{outside}} + \sum \vec{F}(t)_{\text{MS}}$, where $\sum \vec{F}(t)_{\text{outside}}$ is the sum of all of the forces from sources *outside* of the body and $\sum \vec{F}(t)_{\text{MS}}$ is the sum of all of the forces from sources *inside* of the body, i. e., the net musculoskeletal force. Similarly, the net torque can be written as $\sum \vec{\tau}(t) = \sum \vec{\tau}(t)_{\text{outside}} + \sum \vec{\tau}(t)_{\text{MS}}$. Solving eq. (6.1) and eq. (6.2) for the net musculoskeletal force and net musculoskeletal torque, respectively, we have:

$$\sum \vec{F}(t)_{\text{MS}} = m\vec{a}(t) - \sum \vec{F}(t)_{\text{outside}} \tag{6.3}$$

$$\sum \vec{\tau}(t)_{\text{MS}} = \frac{d\vec{L}(t)}{dt} - \sum \vec{\tau}(t)_{\text{outside}} \tag{6.4}$$

The sources of outside forces and torques on a body segment are due to the force of gravity and contact forces such as the ground reaction force or the force due to a partner's hand if the forces are exerted directly on the segment of interest. The force of gravity on the segment always acts vertically downward at the segment's center of mass with a magnitude of $F_g = mg$ where m is the mass of the segment.

For an isolated joint movement like the arm lift with shoulder abduction discussed previously, a single segment rotates around one joint and the net musculoskeletal force and net musculoskeletal torque arise from the muscles, tendons, ligaments, and bones spanning that joint. For a movement involving simultaneous rotations around multiple joints, the analysis is more complex. For example, let's imagine a dancer lifts a leg through développé to a straight leg extension (i. e., simultaneously executing hip flexion and knee extension). If we consider the thigh segment as our system, because the thigh and shank are linked via the knee joint, the motion of the thigh is influenced by the motion of the shank (and vice versa). Muscles spanning both the hip joint (hip flexors and extensors) and knee joint (knee extensors and flexors) contribute to the net torque and net force on the thigh.

Example. Find the net musculoskeletal torque around the shoulder joint as the dancer's lifts the arm with the experimentally measured acceleration in the previous example.

Solution. Since we already determined the net torque as a function of time, we simply subtract the net outside torque to find the net musculoskeletal torque (eq. (6.4)). In this scenario, $\sum \vec{\tau}_{\text{outside}} = \vec{\tau}_g$, where $\vec{\tau}_g$ is the torque due to gravity on the arm segment around the shoulder,

$$\vec{\tau}_g = \vec{r}_{\text{CM}} \times \vec{F}_g = r_{\text{CM}} mg \sin \theta (-\hat{y})$$

where r_{CM} is the position of the center of mass of the arm segment measured longitudinally from the shoulder, m is the mass of the arm segment, and θ is the shoulder abduction angle. Positive $\hat{y}$ values indicate abduction, so the negative torque due to gravity is in the adduction direction. The net musculoskeletal torque is then

$$\sum \vec{\tau}(t)_{\text{MS}} = \sum \vec{\tau}(t) - \vec{\tau}_g = \sum \vec{\tau}(t) + r_{\text{CM}} mg \sin \theta \, \hat{y}$$

We know the longitudinal positions of the upper arm, forearm, and hand: $r_{\text{UA}} = 15.94$ cm, $r_{\text{FA}} = 39.78$ cm, and $r_{\text{H}} = 60.03$ cm. The center of mass of the entire arm segment is found using $r_{\text{CM}} = \frac{\sum m_i r_i}{\sum m_i} = [(1.53 \text{ kg})(15.94 \text{ cm}) + (0.828 \text{ kg})(39.78 \text{ cm}) + (0.336 \text{ kg})(60.03 \text{ cm})]/[2.694 \text{ kg}] = 28.77$ cm. Substituting the numerical values for r_{CM} and m, we have for the net muscle torque as a function of time in the abduction direction

$$\sum \tau(t)_{\text{MS}} = \sum \tau + \left(7.603 \, \frac{\text{N m}}{\text{rad}} \right) \sin \theta \tag{6.5}$$

Combining this with the experimental calculation of overall net torque ($\sum \tau$), we determine the net musculoskeletal torque on the dancer's arm. The results are displayed in Fig. 6.5. We notice that $\sum \tau(t)_{\text{MS}} > 0$ for all times, so for the entire arm lift a net muscle abduction torque is present. We can also find the net MS force on the arm knowing its CM translational acceleration ($\vec{a}_{\text{CM}}$). We first relate $\vec{a}_{\text{CM}}$ to the rotational acceleration of the arm ($\vec{\alpha}$) by noting that $\vec{a}_{\text{CM}} = \vec{\alpha} \times \vec{r}_{\text{CM}}$ where $\vec{r}_{\text{CM}} = r_{\text{CM}}(-\sin \theta \hat{x} - \cos \theta \hat{z})$ (Fig. 6.5). Therefore,

$$\begin{aligned}
\vec{a}_{\text{CM}}(t) &= \alpha \hat{y} \times r_{\text{CM}}(-\sin \theta \hat{x} - \cos \theta \hat{z}) \\
&= \alpha r_{\text{CM}}(-\sin \theta (\hat{y} \times \hat{x}) - \cos \theta (\hat{y} \times \hat{z})) \\
&= \alpha r_{\text{CM}}(\sin \theta \hat{z} - \cos \theta \hat{x})
\end{aligned}$$

This then means that the net x- and z-components of musculoskeletal force on the arm can be found from

$$\begin{aligned}
\sum \vec{F}_{\text{MS}} &= m\vec{a}_{\text{CM}} - \sum \vec{F}_{\text{outside}} \\
\sum F_{\text{MS},x} &= ma_{\text{CM},x} = -m\alpha r_{\text{CM}} \cos \theta \\
\sum F_{\text{MS},z} &= ma_{\text{CM},z} + mg = m(\alpha r_{\text{CM}} \sin \theta + g)
\end{aligned}$$

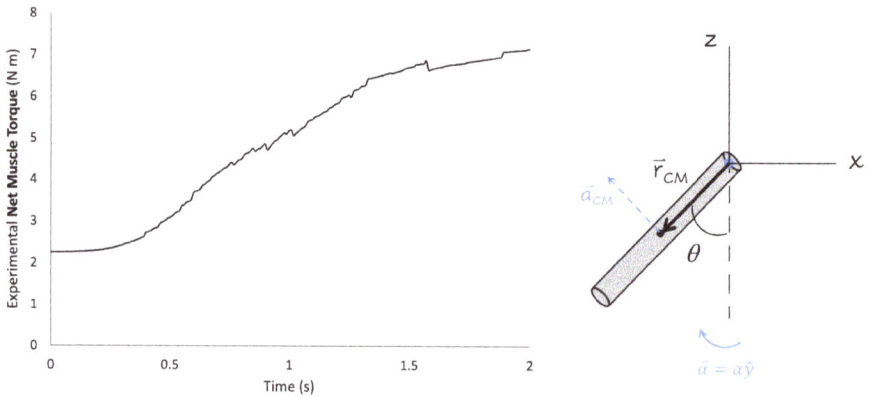

Figure 6.5: Net *musculoskeletal* torque on the arm as a function of time (left). On the right is the global xyz coordinate system (+y into the page) used to determine CM acceleration from rotational acceleration (shoulder abduction) around y.

6.3 Musculoskeletal forces: physical forces inside of the body

What are the individual forces that make up the net musculoskeletal force and torque on a segment? To answer this question, we should remind ourselves of the primary components of the musculoskeletal system discussed in Chapter 2: bones, muscles, tendons, ligaments, and joints. These physical components exert forces, either passively or actively, on a body segment. Let's discuss in more detail the ways in which each of these components can exert force on segments.

6.3.1 Muscle-tendon forces

The forces that we actively control are the muscle forces, which act on a segment via their attachment with tendon. Because the muscle and tendon act on a segment as a unit, the phrases "muscle-tendon force" and "muscle force" are often used synonymously. Muscle forces, like tension forces, can only pull, not push, on a segment.

As learned in Chapter 2, muscle force can vary with time depending on activation level, muscle length and velocity, direction of motion (e. g., flexion, extension, abduction, etc.). There are also multiple muscles that span any given joint. For example, there are 21 muscles that cross the hip joint, but these muscles have different roles in terms of producing motion or stability in different anatomical planes [73]. If a dancer performs a grand battement devant (front kick), the primary muscles that produce hip flexion are the iliacus, psoas, sartorius, and rectus femoris (Fig. 6.6). All of these muscles have attachment sites on the pelvis (origin) and the femur (insertion). If instead the dancer does a grand battement derrière (posterior kick), the primary hip extensors are gluteus maximus, adductor magnus, biceps femoris, semitendinosus, and semimembranosus. There can also be some degree of coactivation during these grand battements, so flexors and extensors may be active simultaneously. Additionally, whether or not these kicks are performed with turnout can affect the activation levels of various muscles utilized.

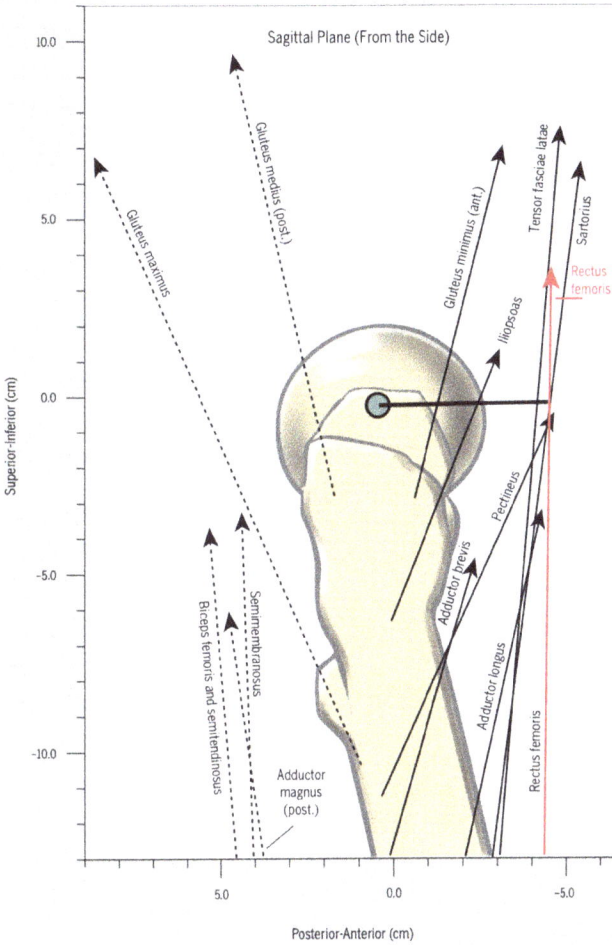

Figure 6.6: Flexor and extensor muscles crossing the hip joint. Sagittal plane view of the femur displaying lines of action of muscle forces (solid arrows are flexors and dashed arrows are extensors). The rectus femoris moment arm from the hip joint center is shown as a bold, black line. Reprinted with permission from [73].

6.3.2 Capsuloligamentous forces

There are forces that hold a joint together, so that, for example, during the grand battement, the femur remains connected with the pelvis and the hip joint does not dislocate. Recall that a human joint is more complex in construction than a simple pin joint between two rods. Ligaments surround the joint to connect the bones, and these ligaments are present both within and outside of the joint capsule. The joint capsule is a space surrounding the connecting bones that is filled with synovial fluid. The fluid itself is very low friction, so we will assume that the torque due to friction is negligible in our analyses. The forces exerted on the segment by the joint capsule and the ligaments crossing a joint can be combined into a single capsuloligamentous force. We also recall from Chapter 2 that the joint capsule and the ligaments remain relatively loose until the limits of the range of motion are approached. At this point, their stiffness increases dramatically. For dancers with hypermobile joints, the ligamentous forces remain small for a larger range of motion.

6.3.3 Bone-on-bone forces

The synovial fluid within a joint capsule acts to provide a low-friction layer between the articulating surfaces. Additionally, cartilage on the ends of the bones serves to protect the bones that are attached at a joint. Higher impact activities can lead to the articulating surfaces in contact to exert large compressive or shear forces on the adjoining bones. Intact cartilage helps protect the bones from damage, but over time the cartilage can become worn, and large, repeated contact forces can lead to wear and tear on the bones' articulating surfaces.

6.4 Creating physical models including musculoskeletal forces

Knowing the magnitudes of specific musculoskeletal forces while dancing would be informative since large or repeated musculoskeletal forces may lead to injury. If our ultimate goal is to determine specific musculoskeletal forces during static or dynamic situations, we need to include them in our physical model (free body diagram) before applying the inverse dynamics approach. To create our physical model, as before we need the mass and inertia of the segment(s) of interest and geometric information about length, orientation, and segment CM position. Additionally, we need information about which muscles are active, where those muscles are attached to the segment and the lines of action of the muscle forces. This requires knowledge of human musculoskeletal anatomy and experience with making reasonable assumptions to create our physical model. We will adopt the following assumptions.
1. Muscle forces only pull.
2. The line of action of a muscle force can be found with a straight line connecting its attachment points (origin and insertion) on adjoining bones. The attachments are determined anatomically from body scans or from data reported in the literature. (This assumption can be flawed if, for example, a muscle must curve around a bony structure to connect the attachment points.)

3. The joint force acts through the axis of rotation of the segment and arises due to the contact force with the adjoining bone.
4. Frictional torque at the joints is negligible.
5. The forces in ligaments are negligible for movements not near the limits of the dancer's range of motion.

In the following sections we will demonstrate this process for movement of the leg segment in a kick. We begin with a single muscle force acting in a static case and build up to dynamic movement involving multiple muscles. The process outlined here can be applied to different body segments and dynamic motions of interest.

Let's consider the grand battement devant. Imagine a dancer utilizes hip flexion to kick the leg anteriorly and that we have measured kinematic data. To start simple, let's assume that only a single muscle force due to the iliopsoas (iliacus and psoas group) is active. We can find the line of action of the muscle from the positions of its origin and insertion as reported in the literature. As the femur rotates in hip flexion, the muscle's insertion point on the femur also rotates around the HJC. If $\vec{r}_i$ is the insertion (femoral attachment) and $\vec{r}_o$ is the origin (pelvic attachment) measured from the HJC, the line of action of force would be in the direction of the separation vector $\vec{r}_o - \vec{r}_i$ (Fig. 6.7). The direction (unit vector) of the *line of action* of the muscle force can be found from

$$\hat{F} = \frac{\vec{r}_o - \vec{r}_i}{\|\vec{r}_o - \vec{r}_i\|} \tag{6.6}$$

Then the torque exerted by that muscle on the bone to which it is inserted is

$$\vec{\tau}_m = \vec{r}_i \times F_m \hat{F} \tag{6.7}$$

where the insertion $\vec{r}_i$ is the position at which the muscle force is applied, measured from the axis of rotation (the hip joint center (HJC) in our example).

In our analysis, we assume the pelvis remains stationary and the femur rotates around the HJC. Thus, we treat the femoral coordinate system as the rotating frame $(x'y'z')$ and the pelvis coordinate system is the static frame (xyz). The common origin of both the pelvis and femur coordinate systems is at the HJC, and the coordinate systems are aligned (with $x = x'$, $y = y'$, and $z = z'$) when in anatomical position. To determine $\hat{F}$, we first express the femoral attachment point in the non-rotating xyz frame. For hip flexion (rotation around the $x = x'$ axis), to transform coordinates from the rotated femur frame to the non-rotated pelvis frame,

$$x_i = x_i'$$
$$y_i = y_i' \cos\theta - z_i' \sin\theta$$
$$z_i = y_i' \sin\theta + z_i' \cos\theta$$

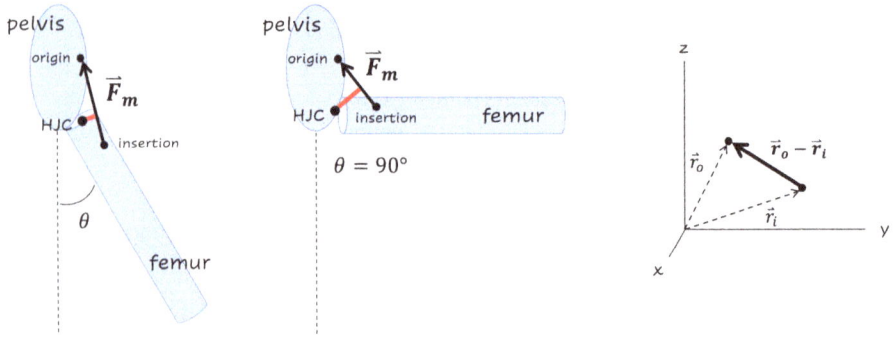

Figure 6.7: The line of action of a muscle force $\vec{F}_m$ is in the direction of a vector connecting its attachment points (e. g., the force that the iliopsas muscle exerts on the femur). The femur and pelvis articulate at the hip joint center (HJC) and a sagittal plane view is shown where the line of action and moment arm of the muscle force (shown in red) change depending on hip flexion angle (θ). The right most image shows the line of action is the direction of the separation vector ($\vec{r}_o - \vec{r}_i$), where $\vec{r}_o$ is the position of the muscle's attachment on the pelvis and $\vec{r}_i$ is the muscle's insertion point on the femur.

So, to express the position of the femoral attachment point as it rotates around the HJC, we have

$$\vec{r}_i = x_i'\hat{x} + (y_i' \cos \theta - z_i' \sin \theta)\hat{y} + (y_i' \sin \theta + z_i' \cos \theta)\hat{z} \tag{6.8}$$

where θ is the hip flexion angle.

6.4.1 Static analysis: single muscle force

Example. Find the line of action of the iliopsoas muscle force on the femur in the sagittal plane when the leg is lifted with 90° of hip flexion.

Solution. In their own local frames, the pelvic and femoral attachments for iliopsoas are at position $\vec{r}_o = (x, y, z) = (0.5, 2.8, 2.4)$ cm and $\vec{r}_i' = (x', y', z') = (1.5, -0.2, -6.1)$ cm [22] measured from the HJC. Since we only wish to find the component of the muscle force in the sagittal plane (yz), we will ignore the x-component. We first find the components of the femoral attachment in the nonrotating frame $\vec{r}_i = (y_i, z_i)$ using eq. (6.8).

$$y_i = -0.2 \cos 90° - (-6.1) \sin 90° = 6.1 \, \text{cm}$$
$$z_i = -0.2 \sin 90° + (-6.1) \cos 90° = -0.2 \, \text{cm}$$
$$\longrightarrow \vec{r}_i = (6.1\hat{y} - 0.2\hat{z}) \, \text{cm}$$

The separation vector $\vec{r}_o - \vec{r}_i = (2.8 - 6.1) \, \text{cm} \, \hat{y} + (2.4 - (-0.2)) \, \text{cm} \, \hat{z} = (-3.3\hat{y} + 2.6\hat{z}) \, \text{cm}$, and we use eq. (6.6) to determine the line of action of the iliopsoas in the yz-plane.

$$\hat{F} = \frac{\vec{r}_o - \vec{r}_i}{\|\vec{r}_o - \vec{r}_i\|}$$

$$= \frac{-3.3\hat{y} + 2.6\hat{z}}{\sqrt{(-3.3)^2 + (2.6)^2}}$$

$$= -0.79\hat{y} + 0.62\hat{z}$$

This means that the iliopsoas line of action makes approximately 38° angle with the line of the femur ($\tan^{-1}(\frac{0.62}{0.79})$, posteriorly and vertically upward) when the leg is held with 90° of hip flexion.

Example. A female dancer (mass = 61.9 kg and height = 173.5 cm) holds her leg at rest parallel to the ground with 90° hip flexion. Assume the iliopsoas is the only muscle group active and find the magnitude of the iliopsoas muscle force ($\|\vec{F}_m\|$) required and the joint force exerted on the femur at the hip.

Solution. A free-body diagram of the external forces on the leg segment in this situation is shown in Fig. 6.8. The force of gravity of the entire leg segment (consisting of the thigh, shank, and foot) acts at its CM and exerts torque in the extension direction ($-\hat{x}$). The iliopsoas muscle force exerts torque (τ_m) in the flexion direction ($+\hat{x}$), and the joint force acts at the axis of rotation (HJC), so it produces zero torque. For the leg to be held at rest,

$$\Sigma \vec{\tau} = 0$$

$$\tau_m - m_{\mathrm{CM}}gr_{\mathrm{CM}}\hat{x} = 0$$

$$\tau_m = m_{\mathrm{CM}}gr_{\mathrm{CM}}\hat{x}$$

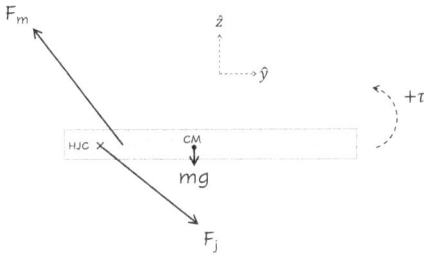

Figure 6.8: Free-body diagram of the forces on the leg segment when held statically with 90° hip flexion. The forces acting are the weight of the leg at its CM, the iliopsoas muscle force F_m and the joint force F_j at the hip joint center. Torque is positive in the flexion direction.

We first find the radial distance of the leg's CM from the hip joint (r_{CM}). From eq. (3.1) and using fractional segment masses from Tab. 3.1 we have

$$r_{\mathrm{CM}} = \frac{m_{th}r_{th} + m_{sh}r_{sh} + m_f r_f}{m_{th} + m_{sh} + m_f}$$

$$= \frac{0.1478r_{th} + 0.0481r_{sh} + 0.0129r_f}{0.1478 + 0.0481 + 0.0129}$$

We find the radial positions of the thigh, shank, and foot (Fig. 6.9) using

$$r_{th} = 0.3612 l_{th} = 0.133\,\text{m}$$
$$r_{sh} = l_{th} + 0.4416 l_{sh} = 0.559\,\text{m}$$
$$r_f = l_{th} + l_{sh} + 0.4014 l_f = 0.893\,\text{m}$$

and inserting these into the expression for r_{CM} we find $r_{CM} = 0.278\,\text{m}$.

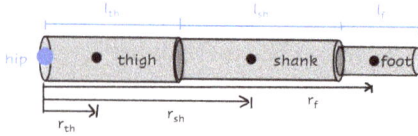

Figure 6.9: Gesture leg segment consisting of the thigh, shank, and foot. Lengths of each segment are denoted by l_{th}, l_{sh}, and l_f, and segment CM positions measured from the hip joint center are r_{th}, r_{sh}, and r_f for the thigh, shank, and foot, respectively.

The total mass of the leg segment is $m = (\frac{m_{th}}{M} + \frac{m_{sh}}{M} + \frac{m_f}{M})M = 0.2088 \cdot 61.9\,\text{kg} = 12.92\,\text{kg}$. This means that the muscle torque required is

$$\tau_m = m_{CM} g r_{CM} = (12.92\,\text{kg})(9.81\,\text{m/s}^2)(0.278\,\text{m})$$
$$= (35.3\,\text{N m})\hat{x}$$

where $+\hat{x}$ is the flexion direction. We then use eq. (6.7) to find the iliopsoas muscle force needed to produce this torque. We use the line of action ($\hat{F}$) and iliopsoas attachment measured from the HJC ($\vec{r}_i$) found in the previous example.

$$\vec{\tau}_m = \vec{r}_i \times F_m \hat{F}$$
$$(35.3\,\text{N m})\hat{x} = (0.061\hat{y} - 0.002\hat{z})\,\text{m} \times F_m(-0.79\hat{y} + 0.62\hat{z})$$
$$= (0.0394\,\text{m})F_m \hat{x}$$
$$\longrightarrow F_m = 895.9\,\text{N}$$

The iliopsoas would have to exert nearly 900 N (equivalent to about 200 pounds) of force to simply hold the leg in a static position with 90° of hip flexion. This greatly exceeds the weight of the leg (approximately 127 N) due to the muscle's small moment arm. In general, muscles are attached relatively close to joint axes of rotation, so they must exert large forces compared to the external loads on a segment.

To find the joint force F_j, we apply the additional requirement for static equilibrium: the net external force on the leg must equal zero.

$$\Sigma F_y = 0 \quad \text{and} \quad \Sigma F_z = 0$$

The iliopsoas muscle force in component form is

$$\vec{F}_m = F_m \hat{F} = (895.9 \text{ N})(-0.79\hat{y} + 0.62\hat{z}) = (-707.8\hat{y} + 555.5\hat{z}) \text{ N}$$

The y-component of the muscle force points in the posterior direction, and the z-component points vertically upward, both of which make sense. We now solve for the y- and z-components of the joint force.

$$\Sigma F_y = 0 \qquad\qquad\qquad \Sigma F_z = 0$$
$$F_{j,y} + F_{m,y} = 0 \qquad\qquad F_{j,z} + F_{m,z} - mg = 0$$
$$F_{j,y} = 707.8 \text{ N} \qquad\qquad F_{j,z} = -428.9 \text{ N}$$

The hip joint structure must exert force anteriorly and vertically down so the leg does not accelerate. As the iliopsoas exerts large force posteriorly and upward, it pulls the femur into the socket of the pelvis (acetabulum), so F_j is the force that the joint socket exerts back on the femur. The magnitude of $F_j = \sqrt{F_{j,y}^2 + F_{j,z}^2} = 827.6 \text{ N} \approx 186 \text{ lbs}$. (Because the force that the joint structure exerts on the femur is an action–reaction pair with the force that the femur exerts on the joint, the term "joint reaction force" may be used in this context. We will, however, stick with the simpler "joint force" terminology.)

6.4.2 Constant velocity analysis: single muscle force

If the leg in the previous example is instead lifted slowly from $0°$ to $90°$ hip flexion, we can use the inverse dynamics process to find F_m of the iliopsoas as a function of flexion angle. If the lift is executed with an approximately constant rotational velocity, then the leg's rotational acceleration $\vec{\alpha} \approx 0$. Like the static case, if rotational acceleration is zero, the net torque and net force on the leg are still zero; however, there are two major physical factors to take into account in this analysis compared to the truly static case. The torque due to gravity varies with hip flexion angle, increasing from zero when $\theta = 0$ to its maximum magnitude when $\theta = 90°$.

$$\tau_g = -mgr_{\text{CM}} \sin \theta \hat{x}$$

The iliopsoas moment arm also varies as a function of θ as its insertion point moves relative to the HJC during hip flexion (Fig. 6.7). The moment arm $r_{i\perp}$ is found with the vector cross product

$$r_{i\perp} = ||\vec{r}_i \times \hat{F}|| \qquad\qquad\qquad (6.9)$$

where the iliopsoas force acts at a position (units = cm) of $\vec{r}_i = (-0.2 \cos \theta + 6.1 \sin \theta)\hat{y} + (-0.2 \sin \theta - 6.1 \cos \theta)\hat{z}$ from the HJC. This is found from eq. (6.8) as in the static example but remains a function of θ. Performing this calculation and a plot of $r_{i\perp}$ versus θ

(Fig. 6.10) reveals that the moment arm ranges from a minimum of less than 2 cm at 0° flexion to a maximum of not quite 4 cm when θ is nearly 90°.

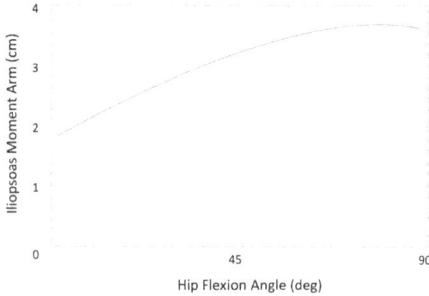

Figure 6.10: Iliopsoas moment arm from the hip joint center as a function of hip flexion angle.

Example. Find the iliopsoas muscle force and the net joint force as a function of angle during an approximately constant rotational velocity leg lift (hip flexion) from 0° to 90°.

Solution. Since $\Sigma\vec{\tau} = 0$, the magnitude of the muscle torque equals the magnitude of the torque due to gravity.

$$\vec{\tau}_m = -\vec{\tau}_g = -(-mgr_{\text{CM}} \sin\theta\hat{x})$$
$$= (35.3\,\text{N m}) \sin\theta\hat{x}$$

And to find the magnitude of the muscle force, we divide torque by moment arm.

$$F_m = \frac{\|\vec{\tau}_m\|}{r_{i\perp}}$$

The magnitude of the iliopsoas muscle force increases as flexion angle increases (Fig. 6.11). The shape of the y- and z-components of the muscle force are due to the line of action ($\hat{F}$) and how it changes with θ. There is a relatively small anterior (y) component of F_m from 0 to approximately 30°, then the y-component increases in the posterior direction as the leg continues lifting from 30° to 90° hip flexion.

The joint force can be found from $\Sigma\vec{F} = 0$, with the same forces acting as did in the static case, but with values now varying with θ.

$$F_{j,y}(\theta) = -F_{m,y}(\theta) \quad \text{and} \quad F_{j,z}(\theta) = mg - F_{m,z}(\theta)$$

Using the values of $F_{m,y}$ and $F_{m,z}$ as functions of θ we find the y- and z-components of the joint force F_j and its magnitude (Fig. 6.12). The initial decrease in F_j is due to the decrease in magnitude in its y- and z-directions. Additionally, since $F_{j,y} = -F_{m,y}$ we observe the y-component of F_j vs. θ to be the mirror reflection of $F_{m,y}$. The z-component of F_j is also

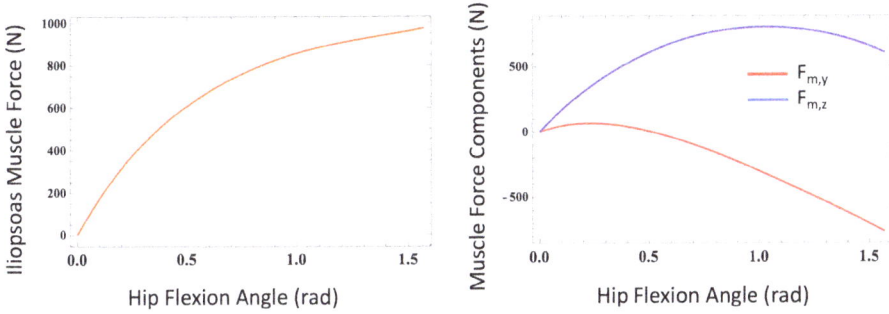

Figure 6.11: Magnitude of iliopsoas muscle force (left) and its anteroposterior (y) and vertical (z) components (right) assuming it is the only muscle force active during a constant velocity leg lift (hip flexion).

a mirror reflection of $F_{m,z}$ but with a vertical shift due to the constant force of gravity ($F_g = 126.8\,\text{N}$), since $F_{j,z} = -F_{m,z} + 126.8\,\text{N}$.

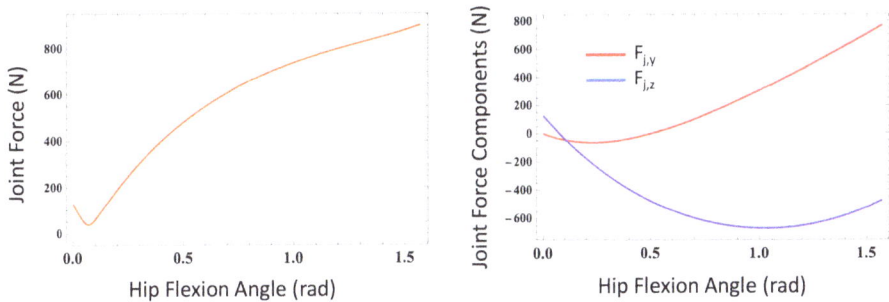

Figure 6.12: Magnitude of hip joint force on the femur (left) and its anteroposterior (y) and vertical (z) components (right) during a constant velocity leg lift (hip flexion) assuming only the iliopsoas muscle is active.

6.4.3 Dynamic analysis: single muscle force

Let's now discuss how to solve for muscle force as a function of angle during a dynamic (accelerating) kick. We must now use measurements of acceleration (a and a_{CM}) and the inertia of the limb. Since the rotation occurs around a symmetry axis of the limb, Newton's second law for rotations and translations is

$$\Sigma \vec{\tau} = I\vec{\alpha} \qquad \Sigma F_y = ma_y \qquad \Sigma F_z = ma_z$$
$$\tau_m - \tau_g = I\alpha \qquad F_{j,y} + F_{m,y} = ma_y \qquad F_{j,z} + F_{m,z} - mg = ma_z$$

The rotational acceleration of the leg around the hip joint and translational acceleration of its CM were measured experimentally for an entire front kick (from $\theta = 0$ to approximately 90° and back down to $\theta = 0$) and data are displayed graphically in Fig. 6.13. To find the total inertia of the leg around the HJC during hip flexion, we use the radii of gyration around segment CMs given in Table 5.1 (r''_x for sagittal plane motion) and the parallel axis theorem, as we did when finding the inertia of the arm around the shoulder.

$$
\begin{aligned}
I &= I_{th} + I_{sh} + I_f \\
&= I_{CM,th} + m_{th}r_{th}^2 + I_{CM,sh} + m_{th}r_{sh}^2 + I_{CM,f} + m_{th}r_f^2 \\
&= 1.9\,\text{kg·m}^2
\end{aligned}
$$

where r_{th}, r_{sh}, and r_f are the radial distances of the thigh, shank, and foot, respectively, from the HJC (Fig. 6.9).

The muscle torque required to kick the leg is

$$
\begin{aligned}
\tau_m &= I\alpha + \tau_g \\
&= I\alpha + mgr_{CM}\sin\theta
\end{aligned}
$$

This muscle torque is plotted in Fig. 6.13. Note that this τ_m would be the total muscle torque required, no matter how many muscles are acting. This same expression can also be applied for any measured α. Calculations such as this using experimentally measured accelerations can be performed with software such as Mathematica. The result for the muscle torque as a function of θ is given in Fig. 6.13. Also displayed are the iliopsoas muscle force and joint force in Figures 6.14 and 6.15.

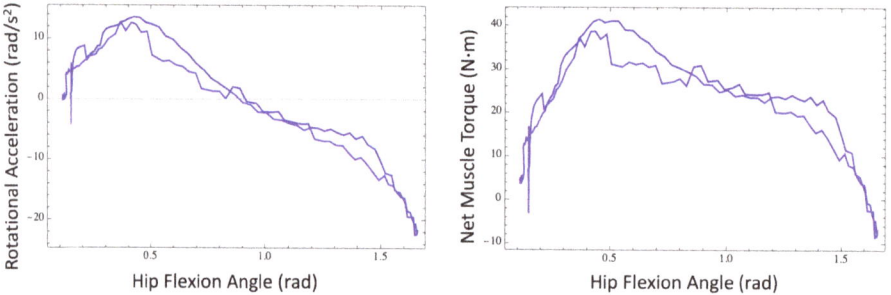

Figure 6.13: Experimentally measured rotational acceleration of the leg around the HJC (left) and the net muscle torque required to produce this acceleration (right). The two nearly overlapping traces are due to the dynamic kick including hip flexion immediately followed by extension back to zero degrees.

We started simple by assuming only one muscle force acts, whereas in reality there are additional primary flexors, other secondary or synergistic muscles activated, as well as possible coactivation of extensors during the kick. In the following section we will see how things quickly get complicated if we add even just one other muscle to our model.

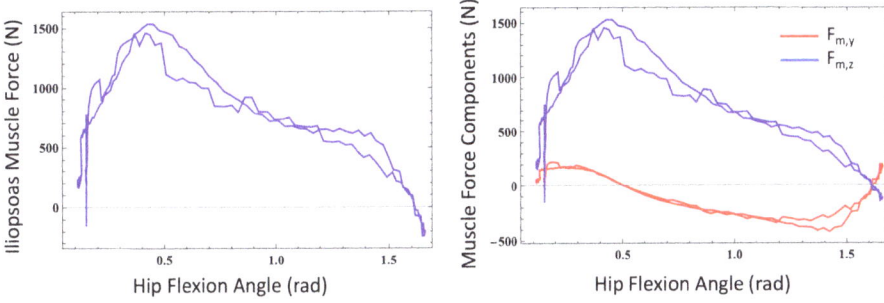

Figure 6.14: Magnitude of iliopsoas muscle force (left) and its anteroposterior (y) and vertical (z) components (right) assuming it is the only muscle force active during a front kick (hip flexion) with the acceleration measured in Fig. 6.13.

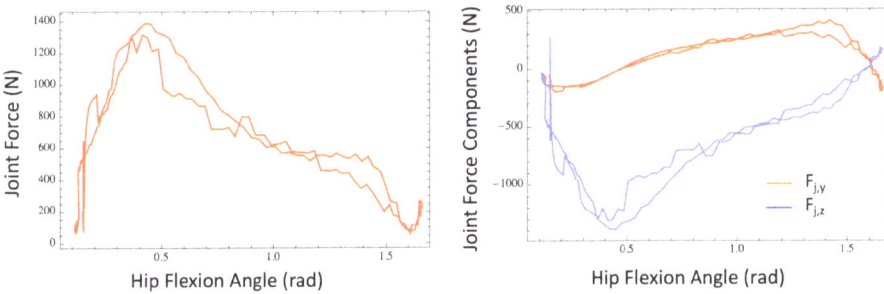

Figure 6.15: Magnitude of hip joint force on the femur (left) and its anteroposterior (y) and vertical (z) components (right) assuming iliopsoas is the only muscle force active during a front kick (hip flexion) with the acceleration measured in Fig. 6.13.

6.5 Multiple muscle forces: the underdetermined system

Let's investigate what happens if we do not simplify our model and assume that all 14 muscles crossing the hip are active during the movement. Each muscle exerts a force $\vec{F}_m$ at a position $\vec{r}_m$ from the axis of rotation (hip joint). Both $\vec{r}_m$ and line of action of $\vec{F}_m$ (and therefore its moment arm $r_{\perp m}$) are known throughout the movement from anatomical information, but the magnitudes of the 14 muscle forces are unknown. Additionally, there is an unknown joint force $\vec{F}_j$ acting at the hip joint (which, as a vector, is really two unknowns (magnitude and direction)). Using eq. (6.4) for the rotational dynamics of the kick, we have

$$\sum_m (\vec{r}_m \times \vec{F}_m) = \frac{d\vec{L}}{dt} - \vec{r}_{\mathrm{CM}} \times \vec{F}_g$$

For motion in a single plane (e. g., sagittal), this becomes

$$\sum_m F_m r_{\perp m} = I\alpha + mgr_{\perp \mathrm{CM}}$$

And the y- and z-components of eq. (6.3) to describe the translational dynamics of the segment's CM become

$$\sum_m F_{m,y} + F_{j,y} = ma_y$$

and

$$\sum_m F_{m,z} + F_{j,z} = ma_z + mg$$

We now have three independent equations and more than 3 unknowns. With only one muscle force in our model, we were able to solve for a single solution because there were exactly 3 unknowns (F_m, $F_{j,y}$, and $F_{j,z}$). Introducing even just one more muscle into our model now leads to a case where there is not a single solution, but an infinite set of possible solutions, unless we place some other criteria on the forces. This means that there is not just one possible way that the muscles can exert their forces to produce the observed movement (acceleration), but infinite possible combinations. This is referred to as the *problem of indeterminacy*.

Our next task is to consider how we might deal with the problem of indeterminacy and reduce our set of combinations of muscle forces to determine which are realistic within human movement. There are several possibilities, which involve introducing additional assumptions or experimental data into our analysis. It is possible to experimentally measure muscle activation (with electromyography) which provides information about which muscles are active and when during dynamic movement. Caution must be taken, however, because it has been shown that the amplitude of an EMG signal (even when normalized to a maximum voluntary contraction (MVC)) does not scale directly to muscle force. Other examples of assumptions one might make are that the active muscles exert force proportional to their cross-sectional area, or that the muscles activate in such a way to optimize an output (e. g., minimize muscle fatigue). These various methods, their use and limitations, are discussed in the following sections.

6.5.1 Muscle scaling

In the method of *muscle scaling*, one of the simplest methods for dealing with the problem of indeterminacy, we assume that the stress in each muscle is constant, i. e., that the force a muscle exerts is proportional to its physiological cross-sectional area. (Physiological cross sectional area differs from standard cross-sectional area as explained later in this section.) The reasoning behind this assumption is that a muscle's maximum strength tends to increase as its physiological cross sectional area increases. We can make conceptual sense of this by considering how muscle fibers are arranged to compose a whole muscle. If fibers are added in parallel (increasing the cross-sectional area of a muscle), it has a similar effect as adding springs in parallel. Two identical springs in parallel, when stretched or compressed the same distance as a single spring, will exert twice as much force. (Adding muscle fibers in series, on the other hand, does not increase the overall force generating capability of a muscle, but increases the total change in length over which a muscle can act.) In an actual muscle, the fibers are not always arranged in line with the longitudinal axis of the muscle and the tendon line. Muscles can be classified by their overall shape and fiber arrangement (Fig. 6.16). In a parallel muscle like the sartorius in the anterior part of the thigh, muscle fibers *are* arranged in parallel. In other muscles (pennate muscles), the fibers make an angle with the longitudinal axis, called the pennation angle. To find the geometric cross-sectional area of an object, we would find the area perpendicular to the longitudinal axis of the object. Instead, *physiological cross sectional area* (PCSA) is the cross-sectional area of a muscle that is perpendicular to its muscle fiber arrangement. Larger pennation angle increases PCSA, so pennation overall increases the force generating capability of a muscle. As pennation increases, however, it decreases the change in length capabilities of a muscle. Rectus femoris is a bipennate muscle, meaning that the muscle fibers are arranged on two sides of a central tendon. If we assume that muscle stress is constant for all muscles, this allows all of the muscle forces to be expressed in terms of just one reference muscle force (eq. (6.10)).

$$\text{If} \quad \text{Stress} = \frac{F_m}{(PCSA)_m} = \text{constant}$$

$$F_{m,j} = \frac{(PCSA_j)}{PCSA_1} F_{m,1} \tag{6.10}$$

where $F_{m,j}$ and $PCSA_j$ are the muscle force and the physiological cross-sectional area corresponding to a muscle, and $F_{m,1}$ and $PCSA_1$ are the muscle force and the physiological-cross sectional area corresponding to the reference muscle. The physiological cross-sectional area of a muscle can be found by dividing the muscle volume by the muscle fiber length, taken from the literature [26]. Now we are able to write all of the muscle forces as some factor of a single muscle force and solve for each individual muscle force.

6.5.1.1 Constant velocity analysis: multiple muscles with force scaling

As an example to demonstrate the process of muscle scaling to solve for multiple muscle forces, we will consider the activation of another primary hip flexor (rectus femoris, in the quadriceps group) in addition to iliopsoas in our analysis of the leg lift with hip flexion. This method can be extended to include more muscles, including any hip extensors that may be coactivated. A discussion of consequences of coactivation can be found in Section 6.5.1.2.

Example. Assume both the rectus femoris and iliopsoas act to rotate the femur relative to the pelvis and execute hip flexion from 0° to 90° slowly with a relatively constant rotational velocity.

(a) Use the reported attachment points of rectus femoris [22] to determine the line of action of its muscle force and its moment arm from the HJC as a function of flexion

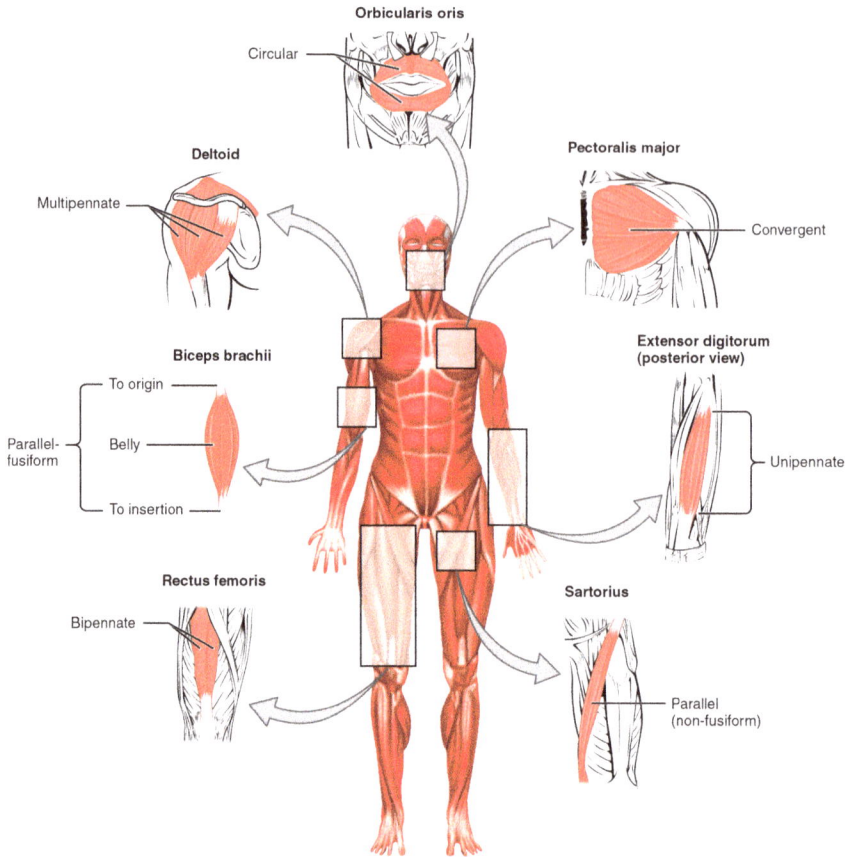

Figure 6.16: "Muscle Shapes and Fiber Alignment" by "OpenStax" is licensed under CC BY. Access for free at https://openstax.org/books/anatomy-and-physiology/pages/1-introduction.

angle. The origin of rectus femoris on the pelvis ($\vec{r}_{o,RF}$) and its insertion on the femur (in the femur's rotating frame ($\vec{r}'_{i,RF}$)) are

$$\vec{r}_{o,RF} = (4.3\hat{y} + 3.7\hat{z}) \text{ cm}$$
$$\vec{r}'_{i,RF} = (4.3\hat{y}' + (-41.5)\hat{z}') \text{ cm}$$

(b) Assuming each muscle maintains an equal amount of stress, use muscle scaling to solve for iliopsoas muscle force ($F_{m,I}$) and rectus femoris muscle force ($F_{m,RF}$) as functions of θ. Physiological cross-sectional areas of the muscles are $PCSA_{RF} = 9.20 \text{ cm}^2$ and $PCSA_I = 12.52 \text{ cm}^2$.

Solution. (a) From eq. (6.8), the position of rectus femoris insertion point from the HJC in the non-rotating pelvis frame as a function of flexion angle is

$$\vec{r}_{i,RF} = \left[(4.3\cos\theta + 41.5\sin\theta)\hat{y} + (4.3\sin\theta - 41.5\cos\theta)\hat{z} \right] \text{ cm}$$

The line of action of rectus femoris ($\hat{F}_{RF}$) and its moment arm ($r_{\perp RF}$) are found with eq. (6.6) and eq. (6.9) as was done for iliopsoas. Rectus femoris moment arm as a function of hip flexion angle is plotted in Fig. 6.17.

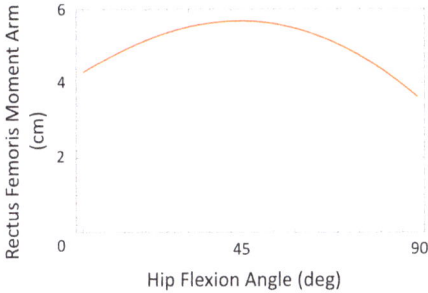

Figure 6.17: Rectus femoris moment arm as a function of hip flexion angle.

(b) For a constant rotational velocity front leg lift engaging the iliopsoas and rectus femoris as hip flexors,

$$\Sigma \vec{\tau} = 0$$

$$-\tau_g \hat{x} + \tau_{m,I} \hat{x} + \tau_{m,RF} \hat{x} = 0$$

$$-(35.3 \text{ N m}) \sin \theta + F_{m,I} r_{\perp,I} + F_{m,RF} r_{\perp,RF} = 0$$

With the muscle scaling substitution that $F_{m,RF} = (\frac{9.2}{12.52}) F_{m,I}$ from eq. (6.10), we have

$$F_{m,I} r_{\perp,I} + \left(\frac{9.2}{12.52} \right) F_{m,I} r_{\perp,RF} = (35.3 \text{ N m}) \sin \theta$$

which can now be solved for the single unknown $F_{m,I}$. Once $F_{m,I}$ is known, we can also solve for $F_{m,RF} = (\frac{9.2}{12.52}) F_{m,I}$. The results are plotted in Fig. 6.18.

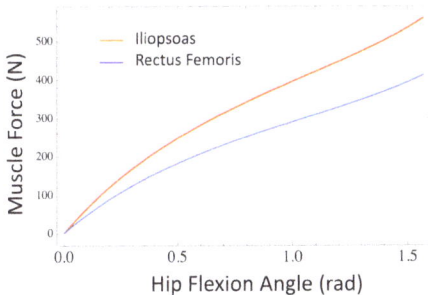

Figure 6.18: Magnitudes of iliopsoas and rectus femoris muscle forces assuming they are the two muscles active and muscle force scales with PCSA during a constant velocity leg lift (hip flexion).

We observe the same overall shape of the muscle force curves for iliopsoas and rectus femoris that we found for iliopsoas alone (Fig. 6.11). Muscle force increases as flexion angle increases from 0° to 90°, with $F_{m,RF}$ always 73.5 % of $F_{m,I}$ as specified by the muscle scaling. Due to the force sharing of each of these flexors, neither reaches as great of a magnitude as the iliopsoas when acting alone (obtaining maximum values of roughly half or less than iliopsoas alone (max of 1000 N)).

6.5.1.2 Coactivation

Let's investigate what happens if the muscles that are simultaneously active across a joint do not produce torque in the same direction, but are oppositely directed torques. Simultaneous activation of agonist and antagonist muscles is termed *coactivation*. In our hip flexion example, this would occur if hip extensor muscles such as the gluteus maximus and biceps femoris (hamstring) were activated along with the primary hip flexors. Let's consider another example of an agonist/antagonist pair: the biceps and triceps muscles crossing the elbow joint.

Example. Imagine a dancer holds the lower arm (forearm plus hand) at rest with 90° of elbow flexion and another dancer pushes down on their hand with an external force. The mass of the lower arm is 1.5 kg and its center of mass is located 14 cm from the elbow joint center (EJC). A downward external force of 65 N also acts at the hand, 30 cm from the EJC. (a) Assume the biceps is the only muscle active, it acts vertically upward, and its moment arm from the EJC is $r_{\perp B} = 3.8$ cm. Find the muscle force required and the joint force at the elbow in this situation. (b) Now assume the triceps and biceps coactivate, that the triceps exerts 30 % as much force as the biceps ($F_T = 0.3F_B$), and its moment arm is $r_{\perp T} = 1.3$ cm. Find the magnitudes of both muscle forces and the joint force during coactivation.

Solution. Free-body diagrams of the two cases are shown in Fig. 6.19. In both cases, the arm is held at rest so the net force and net torque on the arm segment (forearm plus hand) are zero. To find the biceps muscle force when there is no force in the triceps (a), we have

$$\Sigma \vec{\tau} = 0$$
$$\tau_B - \tau_g - \tau_{F_{ext}} = 0$$
$$F_B r_{\perp B} = m_a g r_{\perp a} + F_{ext} r_{\perp ext}$$
$$= (1.5\,\text{kg})(9.8\,\text{m/s}^2)(0.14\,\text{m}) + (65\,\text{N})(0.3\,\text{m}) = 21.56\,\text{N m}$$
$$F_B = \frac{21.56\,\text{N m}}{0.038\,\text{m}} = 567\,\text{N}$$

Then, we find the joint force by knowing that the net vertical force must be zero.

$$\Sigma F_z = 0$$

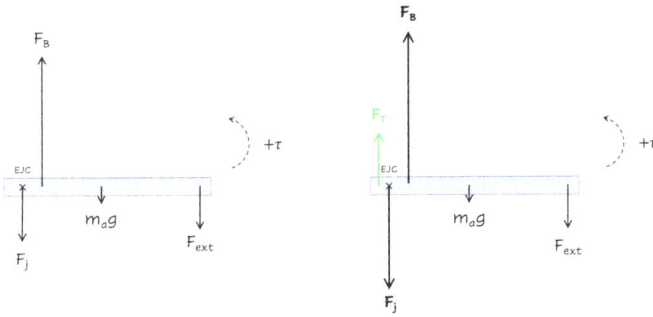

Figure 6.19: Free-body diagrams (not to scale) of the forces on the forearm + hand segment. Only the biceps muscle (F_B) is active on the left. The triceps coactivates (F_T) on the right, and as a consequence both the biceps force and the joint force (F_j) at the elbow joint center (EJC) increase.

$$F_j + F_B - m_a g - F_{\text{ext}} = 0$$
$$F_j = m_a g + F_{\text{ext}} - F_B$$
$$= 14.7\,\text{N} + 65\,\text{N} - 567\,\text{N} = -487\,\text{N}\hat{z}$$

The negative sign indicates that the joint force acts vertically downward. The joint force is a consequence of the large upward biceps force that compresses the forearm segment into the adjoining upper arm segment at the elbow joint. If the joint force were not present, the forearm would accelerate upward under the action of the biceps force. If the triceps coactivates (b), it exerts an upward force on the forearm on the opposite side of the elbow joint. It is an extensor muscle, and its torque is in the $-\hat{x}$-direction like the torques due to gravity and the external force acting at the hand. We again solve for the muscle forces and joint force in this static situation.

$$\Sigma \vec{\tau} = 0$$
$$\tau_B - \tau_T - \tau_g - \tau_{F_{\text{ext}}} = 0$$
$$F_B r_{\perp B} - F_T r_{\perp T} = 21.56\,\text{N m}$$

We see that the biceps force will need to be larger than it was without the coactivation since the triceps produces an additional negative (extension) torque. To solve for F_B, we make the muscle scaling substitution that $F_T = 0.3F_B$ and use the moment arms given in the problem.

$$F_B(0.038\,\text{m}) - (0.3F_B)(0.013\,\text{m}) = 21.56\,\text{N m}$$
$$F_B(0.0341\,\text{m}) = 21.56\,\text{N m}$$
$$F_B = 632\,\text{N}$$

The biceps muscle force increased as expected, and is about 10 % larger than the case without coactivation. Solving for triceps muscle force, we have

$$F_T = 0.3F_B = 0.3(632\,\text{N}) = 190\,\text{N}$$

We now find the joint force during coactivation.

$$\Sigma F_z = 0$$
$$F_j + F_T + F_B - m_a g - F_{\text{ext}} = 0$$
$$F_j = m_a g + F_{\text{ext}} - F_T - F_B$$
$$= (14.7 + 65 - 632 - 190)\,\text{N} = -742\,\text{N}\hat{z}$$

The downward joint force also increased due to the additional upward force of the triceps pulling the lower arm into the elbow joint. Notice that the joint force increased by more than 50 % even with this seemingly slight coactivation. One of the main consequences of coactivation is that it increases the joint forces between segments. By pulling the adjoining segments together with forces on both sides of a joint, coactivation increases joint stability, but at the consequence of larger compressive forces between bones. Because of this, coactivation may increase injury risk, particularly during dynamic tasks with large loads (e. g., jump landings).

6.5.1.3 Limitations of muscle scaling

The simplicity of the muscle scaling approach adds to its appeal as a tool to address the problem of indeterminacy. If we have basic anatomical measurements of moment arms and muscle PCSAs, we can solve for the muscle forces in any number of muscles spanning a joint. We chose to focus on simpler examples with two muscle forces acting, but we can use this method to solve for any number of muscle forces.

There is, however, a major limitation to muscle scaling. While it may feel reasonable that the force a muscle exerts at any time is proportional to its PCSA, that assumption is flawed. Not all muscles spanning a joint are necessarily activated at the same time or to the same magnitude during static and dynamic movements. Research measuring muscle forces directly with tendon buckles has confirmed this flaw in the muscle scaling approach. How much force a given muscle exerts depends upon multiple factors such as the muscle's length and velocity (recall the force–length and force–velocity relationships discussed in Chapter 2) as well as the composition of muscle fibers and their neural activation.

Although muscle scaling is overly simplistic, it is useful as an exercise in learning problem solving approaches to the problem of indeterminacy. If forces do not necessarily scale at some constant value, however, how else might they work together to execute a movement? Perhaps the muscles are activated in such a way to minimize total muscle force exerted (which would minimize joint forces)? Or to minimize muscle fatigue? The next section discusses optimization methods for solving for multiple muscle forces.

6.5.2 Optimization methods

In our inverse dynamics problem, we are faced with the task of solving for more un-knowns than we have equations if more than a single muscle force is assumed to act. This means that there may be an infinite number of combinations of muscle forces that can satisfy the equations of motion of the body segment of interest and produce the desired kinematic results. But what combinations are actually used by individuals or dancers? One way of dealing with this issue is to assume that the actual combination of forces (their magnitudes and timing) minimize some *cost function*, ζ, which can be expressed as a function of the unknown forces ($\zeta = f(F_{m,1}, F_{m,2}, \ldots)$). One simple choice of a cost function could be the total force exerted by all muscles.

$$\zeta_A = F_{m,1} + F_{m,2} + \cdots + F_{m,N} = \sum_{i=1}^{N} F_{m,i} \tag{6.11}$$

where N is the number of muscles in the model. So the solution would find the combi-nation of muscle forces that minimizes ζ_A. Another example of cost function is the total muscle stress, ζ_B.

$$\zeta_B = \sum_{i=1}^{N} \frac{F_{m,i}}{PCSA_i} \tag{6.12}$$

Minimizing ζ_B would produce the combination of muscle forces that minimizes the to-tal stress in all of the muscles. One other cost function that has been suggested in the literature [80] is the muscle fatigue function, ζ_C.

$$\zeta_C = \sum_{i=1}^{N} \left(\frac{F_{m,i}}{PCSA_i} \right)^3 \tag{6.13}$$

and minimizing ζ_C would lead to the combination of forces that minimize total muscle fatigue. There are other possibilities for cost functions, but let's apply these to our hip flexion example with iliopsoas and rectus femoris so we can demonstrate their use and how to interpret results. In the following examples, we use the same kinematic data collected from the dancer performing the front kick as previous sections. Then we will solve for the iliopsoas and rectus femoris muscle forces minimizing the different cost functions and compare the results. All of the cost functions will be subject to the equality constraint

$$F_{m,I} r_{\perp I} + F_{m,RF} r_{\perp RF} = (35.3 \, \text{N m}) \sin \theta + (1.9 \, \text{kg m}^2) \alpha$$

which is simply $\Sigma \vec{\tau} = I \vec{\alpha}$ for the kick.

We also know that the muscle forces can not be negative (this would mean that the muscle forces could push on the leg and exert torque in the extension direction, which is not physically possible), and the muscle must have some upper limit based in its overall strength. So we also subject our cost function to the inequality constraints

$$0 \leq F_{m,I} < F_{m,I,\max}$$
$$0 \leq F_{m,RF} < F_{m,RF,\max}$$

The upper limit can be obtained by knowing maximum possible muscles stress. Research suggests this value varies between around 30–100 N/cm^2, but these values are not necessarily constant. If we chose to use the larger reported maximum stress, this would impose upper limits of $F_{m,I,\max} = 1252$ N and $F_{m,RF,\max} = 920$ N.

6.5.2.1 Minimizing total muscle force: dynamic front kick with two muscle forces

When minimizing total muscle force (ζ_A), iliopsoas force remains at zero for nearly the entire kick, while only rectus femoris acts (Fig. 6.20). We can reason why this happens by recognizing that the moment arm of the rectus femoris is greater than that of the iliopsoas for nearly the entire hip flexion range, so less force is required to exert the same torque at a greater moment arm. Only once the kick reaches (and slightly surpasses) 90° hip flexion, does the iliopsoas moment arm become greater, and its muscle force switches on, while rectus femoris switches off. These results allow us to see that minimizing cost function ζ_A will lead to this on/off switching of muscles, depending on which moment arm is greatest. Muscle activation does not in general switch back and forth between just the muscle with the greatest moment arms at any time, so ζ_A is not in general an effective cost function.

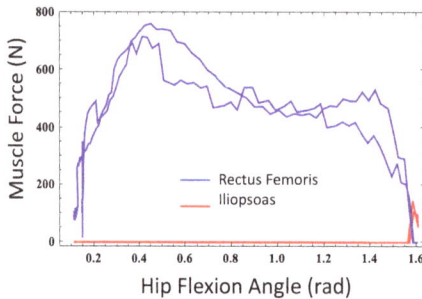

Figure 6.20: Magnitudes of iliopsoas (red) and rectus femoris (blue) muscle forces that would minimize total muscle force required during a dynamic front kick (assuming they are the only two muscles acting).

6.5.2.2 Minimizing total muscle stress: dynamic front kick with two muscle forces

When minimizing total muscle stress (ζ_B), we again obtain an on/off switching between only rectus femoris acting for low degrees of hip flexion, and then only iliopsoas acting for greater degrees of hip flexion (Fig. 6.21). The switch happens earlier in the motion (at around 70° hip flexion instead of 90°), due to the muscle force being scaled to its PCSA.

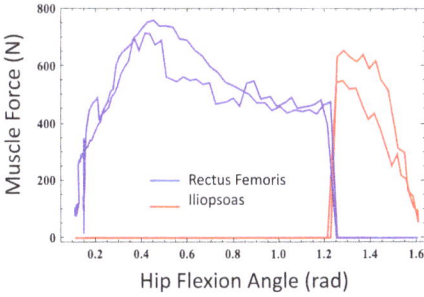

Figure 6.21: Magnitudes of iliopsoas (red) and rectus femoris (blue) muscle forces that would minimize total muscle stress during a dynamic front kick (assuming they are the only two muscles acting).

6.5.2.3 Minimizing muscle fatigue

Finally, minimizing muscle fatigue (ζ_C) yields the muscle force results in Fig. 6.22. These results are similar to those obtained with a muscle scaling approach. It makes sense that to reduce muscle fatigue, both muscles would be engaged to share the load. We observe that rectus femoris muscle force is less than iliopsoas, due to its smaller PCSA. In the end, a variety of muscle activation patterns and combinations of muscle forces can produce exactly the same net joint torque and kinematics. A movement such as this kick can look exactly the same to an outside observer, but the motor control and precise musculoskeletal forces involved may differ. Different individuals may activate the muscles differently depending on a variety of factors, including muscle strength imbalances, training, and whether or not they have an injury. For example, a study on grand battement devant (front kick) found that muscle use (measured with EMG) varied depending on a number of factors, including whether the kicks were performed at the barre, in the center, or traveling, as well as which leg a dancer used and their level of training [51]. It is useful to gather EMG data to inform musculoskeletal modeling. Neither EMG alone nor inverse dynamics alone can give us precise information about magnitudes of musculoskeletal forces within the body. In spite of this, inverse dynamics can still be useful as a tool to estimate musculoskeletal forces and make reasonable recommendations to reduce injury risk if those forces are found to be particularly large.

6.5.3 Electromyography

As we know from Chapter 2, muscles become activated when stimulated by electrical signals from motor neurons. *Electromyography* (EMG) is a procedure used to quantify the level of muscle activity by measuring the potential difference between parts of the muscle. The potential difference can be measured across two parts of the same muscle or between one part of the muscle and a reference or ground electrode. We can imagine if we were able to insert the electrodes of a voltmeter inside of the muscle, we could measure the potential difference across muscle fibers. This is essentially what intramuscular

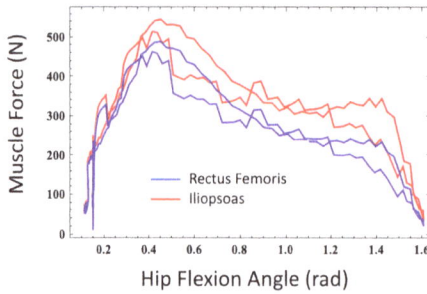

Figure 6.22: Magnitudes of iliopsoas (red) and rectus femoris (blue) muscle forces that would minimize total muscle fatigue during a dynamic front kick (assuming they are the only two muscles acting).

electromyography does. *Intramuscular EMG* uses needles or fine wires inserted directly into the muscle. Some intramuscular EMG electrodes are so small that they can measure the potential difference in a single muscle fiber. EMG electrodes do not need to be inserted directly into the muscle to obtain a signal, however, and electrical activity can be measured by placing electrodes on the skin above the muscle of interest in *surface EMG*. Because it is noninvasive, surface EMG is widely used in human biomechanics research. In this section, we will cover some of the important basics to understand what exactly EMG measures and examples of how it has been used in biomechanics of dance research.

Recall that an action potential (AP) is a single electrical signal from a motor neuron to the muscle fibers it activates. When many APs are sent from the neuron to the muscle fibers in its motor unit, the resulting signal is the summation of those many action potentials. Surface EMG does not isolate a single motor unit, but measures the activity of many motor units that are simultaneously active. How many motor units the EMG includes depends on the placement of the EMG electrodes. The resulting EMG signal is the summation of the many APs of the multiple motor units (hundreds or thousands of APs). Even still, the voltage of this signal is relatively small, so it is amplified prior to analyzing. An example of this type of interference EMG signal (also commonly referred to as raw EMG) is shown in Fig. 6.23. Since APs, and therefore the raw EMG, contain both positive and negative values of voltage, if we wanted to quantify the amount of muscle activation, taking an average of the raw EMG would not be very useful since it's average is approximately zero. To process the EMG signal, it is often rectified (taking the absolute value of the signal) and then low-pass filtered or integrated to remove its high-frequency content. We can think of the integrated signal as the envelope of the rectified EMG signal, and it provides information about the active state of the muscle. Some studies have found a relationship between the amplitude of the integrated EMG curve and muscle force, but only for isometric activation [27].

The exact number of millivolts in an EMG signal is essentially meaningless. Among other things, it depends on the person, the impedance of the tissue surrounding the

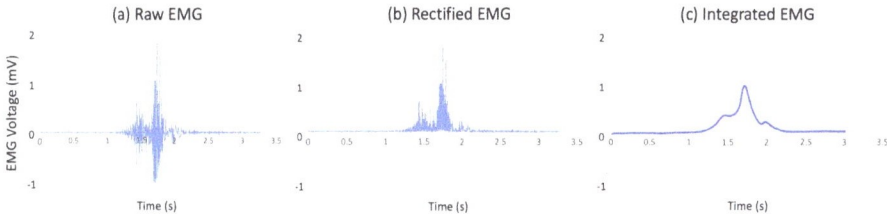

Figure 6.23: (a) Raw EMG signal (b) Rectified EMG and (c) Integrated EMG measured with surface EMG of the rectus abdominus muscle.

muscle, the exact placement of the electrodes on the skin above the muscle, and the amplification of the signal. What is much more meaningful is if we compare the measured EMG signal relative to some reference EMG signal. Very often, this reference EMG signal is one that is recorded during the maximum voluntary (isometric) contraction (MVC) of the muscle. The normalized EMG signal is then given as a percentage of MVC.

It must be emphasized that while EMG can and has been used to estimate muscle force, EMG by itself is not a measure of muscle force and should not in general be used as a substitute for muscle force. Studies have shown that EMG amplitude is not in general directly proportional to muscle force (measured directly with a tendon buckle) [42]. There are several reasons why. One reason is that the amount of muscle force generated depends on not only the muscle's activation but its state (length and velocity) when its activated. Additionally, there is an electromechanical delay in the amount of time between when a muscle is activated and when it begins to produce force.

Since the relationship between EMG and muscle force breaks down during dynamic situations, what can EMG be used for when muscles are activated eccentrically or concentrically as the body moves? The integrated EMG signal carries information about the relative amplitude or size of the muscle activation. EMG amplitude can be used to determine which muscles are turned "on" or turned "off" (and to what degree) at what points in time during movements. For example, one study measured EMG activity in several muscles of the legs during ballet grand plié (knee flexion) and compared activation patterns between ballet dancers and contemporary dancers [100]. Analysis of the EMG activity enabled the researchers to find differences in muscle activity between the dancers of the different styles (Fig. 6.24). Additionally, the frequency content of the EMG signal can be analyzed (e. g., with Fast Fourier Transform). Studies have shown that the frequency content of the signal shifts to being composed of lower frequencies when muscles are fatigued.

EMG can be used to investigate muscle coordination patterns during complex movements (like a leap or jeté), and studies have found patterns in EMG showing that muscles work synergistically to produce complex movements [60]. One study used EMG to compare activation of muscles in the quadriceps group (vastus medialis oblique and vastus lateralis) during sautés and pliés executed with and without turnout [104], and found that both muscles were significantly more activated when using turnout compared to

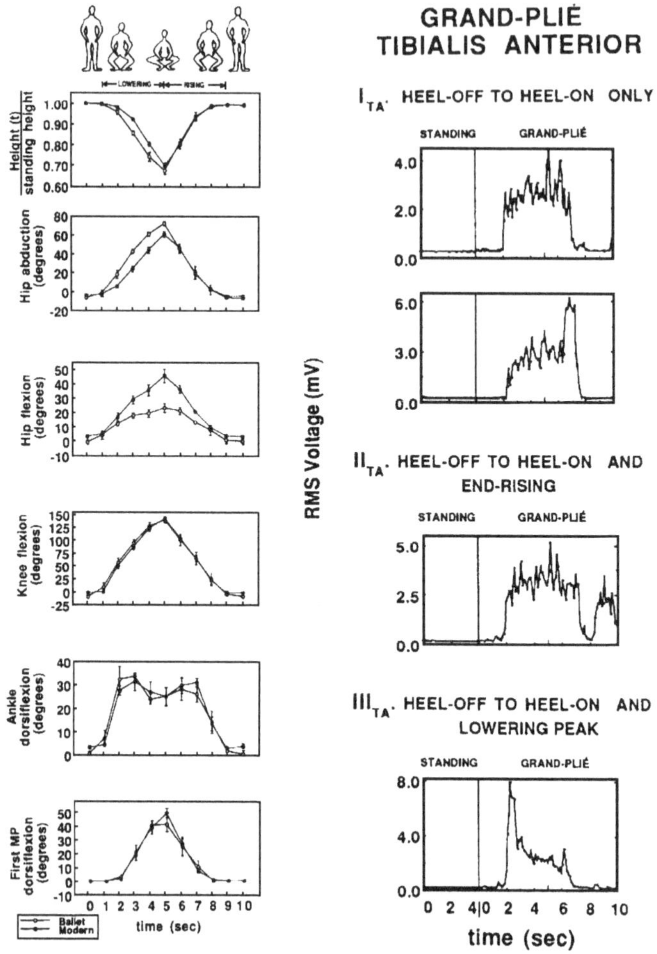

Figure 6.24: Grand plié joint kinematics (left panel) and representative types (I_{TA}, II_{TA}, and III_{TA}) of tibialis anterior (TA) EMG RMS voltage traces (right panel). In this study, all grand pliés showed TA muscle activity from heel off (lowering phase) to heel on (rising phase), but ballet and modern dancers differed significantly in TA activity. The majority of grand pliés performed by ballet dancers exhibited a second peak of TA activity at the very end of the rising phase (Type II_{TA}), whereas modern dancers did not (Type I_{TA}). Reprinted with permission from Wolters Kluwer Health, Inc. [100].

parallel in both ballet and contemporary dancers. Studies have also found differences between dancers and non-dancers in muscle synergies by studying surface EMG measurements from leg and trunk muscles during walking [87].

EMG data can also provide information relevant to injuries. One study compared dancers with and without ankle injury during jump landings with turnout [59]. EMG activity demonstrated that dancers with ankle injury used greater coactivation and greater tibialis anterior (TA) activity, providing more ankle joint stability in a "load avoidance strategy." Another study measured EMG activation in the leg muscles and GRFs on the knees and hips with a force plate during choreographed falls by contemporary dancers [81]. They compared results from two falling techniques (anteriorly and laterally directed falls) finding that the knee contact force decreased and lower extremity muscle activity increased for the laterally directed falls. Their results lead to greater insight into what choreographed falling techniques may be safest for dancers.

Surface EMG does have limitations. It cannot be used to measure activity of deep muscles in which there is another muscle between the muscle of interest and the skin. The surface EMG detects electrical signals from any muscle (or other nearby source), so care must be taken to ensure proper placement of electrodes to measure only the muscle desired. Since the signal depends on the number of muscle fibers the electrodes cross and which motor units are being activated, different electrode placement for the same muscle can produce a different overall signal. This is one reason why it is difficult to directly compare EMG amplitudes from person to person or even within the same person if the electrodes have been moved (e. g., data taken on different days).

6.5.4 Musculoskeletal simulations

Another method for dealing with the problem of indeterminacy is to execute a computational *musculoskeletal simulation*. Simulations can be carried out using open-source software, such as OpenSim, and computational models can include neural and sensory information, information that was missing in the models and inverse dynamics processes previously discussed in this chapter. The simulations can use kinematics as inputs and employ inverse dynamics to solve for torques and forces, or a forward dynamics approach. In the forward dynamics approach, the simulation can use neural information, either from a model or experimental (EMG) data, and a muscle-tendon model that includes force–velocity and force–length properties of muscles to generate muscle forces and simulate the kinematics that would be produced. The kinematics that are generated from the forward dynamics can be compared to experimentally measured kinematics. Fig. 6.25 is a schematic representation of the forward dynamics process. For more information about musculoskeletal modeling, see, for example, [90].

6.6 Musculoskeletal forces in dance and implications for training and dance injury

In spite of its challenges and limitations, inverse dynamics estimates of musculoskeletal forces while dancing can lead to useful information in regards to injury risk or that may impact training. If large muscle or joint forces are found during movements executed with certain kinematics, it could indicate the need to modify technique for injury prevention. Comparisons of muscle and joint kinetics in movements executed by different

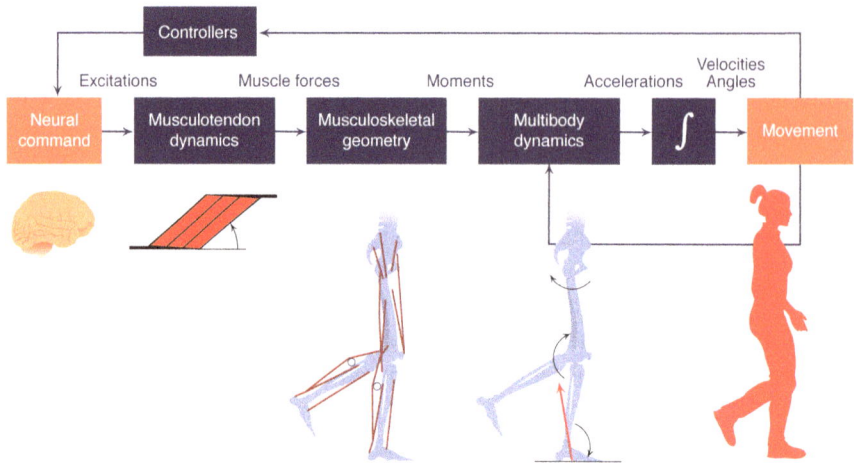

Figure 6.25: Schematic representation of a forward dynamics analysis in a musculoskeletal simulation with OpenSim. Reprinted with permission from [90].

populations (e. g., dancers and non-dancers, males and females, injured and non-injured dancers), and in movements executed before and after fatigue, can also lead to greater knowledge surrounding injury risk. Several examples of research studies from the literature will be discussed to demonstrate some ways that inverse dynamics has been used in biomechanics research in dance.

If musculoskeletal forces are estimated to be large during certain dance movements, it may indicate that those movements pose an increased risk of injury. Even if the forces are not great enough to lead to an acute injury, if sustained during many repetitions over time, they may lead to an overuse injury. Lower extremity net joint torques and forces were measured in a variety of standard tap dance movements, and found to be relatively small when compared to other activities [68]. Mean peak joint forces were all less than two times body weight. As we know, *net* joint torques and forces do not provide information about specific physical forces within the body (e. g., individual muscle force), but these results suggest that tap is a low impact activity and could explain the relatively low rate of injury in tap dance. On the other hand, a study of the Irish dance "rock step" found large forces in the Achilles tendon with ankle joint contact force of 14 times body weight [91]. In the rock step, the dancer stands on the balls of the feet and rocks side to side, inverting one ankle and everting the other in sync. In doing so the dancer rocks from one big toe (hallux) to the other. The researchers suggested that these large forces indicate potential for injury and urge caution when performing the rock step. Studies have also found an increase in lower extremity joint forces and torques as 2D dance jump distance increases. Greater quadriceps force was found with the increased jump distance, and led to an increase in knee joint axial force (measuring up to 14 times body weight) [94].

Another study estimated compressive forces in the lumbar spine when dancers performed vertical movements while attached to a harness [107]. In their musculoskeletal model, researchers used the product of muscle activation (normalized EMG), muscle cross-sectional area from the literature and scaled to the dancers in their study, and muscle stress set to a constant $47.4\,\text{N/cm}^2$. The largest spinal compressive forces were found in body positions in which the dancer's body was mostly vertical (either upside down or right side up) while they were moving the trunk into a hyperextended position. The researchers found that while in the inversion position, dancers utilized coactivation of the abdominals and low back muscles to increase trunk stability. This highlights the tradeoff discussed previously between stability at a joint and joint forces.

Multiple studies have found changes in kinematics and kinetics of jump landings with fatigue [1]. One study found changes in joint kinematics, greater ankle joint force, and greater knee flexion torque post fatigue during landings from an Irish dance jump called a leap over [106]. Another study compared dancers to other athletes and found that both groups exhibited kinematic changes (e. g., increased trunk flexion) and kinetic changes (e. g. increased peak knee valgus moment (pulling knees inward)) post fatigue, and that these types of changes increase the risk of ACL injury [62]. They also found that dancers were more resistant to fatigue than the other athletes, which may explain why dancers experience less ACL injuries than other athletes.

While we may not have the technology to directly measure individual muscle or joint forces while dancing, studies such as these can help inform the dancer at the beginning of the chapter and others in the dance community about certain factors or techniques that may increase injury risk in dance.

7 Energy

While playing tennis with a friend, a dancer becomes winded and needs a break. She gets a drink of water and wonders, "Why do I feel so out of breath? I thought I was in good shape from all my dance classes." While back on the court, she pays attention to what her body does and thinks about her physics class. She serves the ball, then runs to one side of the court to return her friend's shot, and finally sprints forward hitting a volley over the net to win the point. In physics she is learning about work and energy and thinks to herself, "If I started at rest before serving the ball, then I ended at rest after hitting my last shot, my total change in kinetic energy was zero. Doesn't that mean the total work done was zero?! But I sure felt my muscles working hard!" She also realizes that her total change in gravitational potential energy was zero since the tennis court is flat and she started and ended with her center of mass at the same height. "Energy is confusing," she thinks. Then she remembered her physics professor explaining that the combination of kinetic energy and potential energy was called mechanical energy, but that there were also other forms of energy like thermal energy and chemical energy. She resolves to ask her professor about it, but for now she'll get her mind back in the tennis game.

This dancer is certainly not alone in finding energy confusing. As scalar quantities, work and energy can be mathematically simpler than forces and torques, however, energy is a conceptually challenging topic. For one thing, it is very difficult to give a satisfying definition of energy. Additionally, students are told that energy is conserved, but then sometimes it's not. The law of conservation of energy is a fundamental physical principle, and states that the total energy of an isolated system remains constant. An *isolated system* is one in which no energy is exchanged between the system and environment (anything not included in the system). The entire universe is an isolated system, so the universe's total energy remains fixed. When analyzing human movement, our system will not include the entire universe however, so we can not necessarily assume the system is isolated for all cases.

Even after defining a system, the forms that energy takes can constantly change within that system. Keeping track of all the types of energy within a system and how much energy is transferred into or out of the system becomes one big accounting problem. Students of introductory physics, like the tennis playing dancer, are most familiar with two types of energy: kinetic (energy of motion) and potential (stored energy), and the total *mechanical energy* of a system is the sum of its kinetic and potential energies ($E_{mech} = K + U$).

Introductory physics students may also encounter the idea of a more mysterious "internal energy." For example, if a book falls toward the ground, the kinetic energy of the book-Earth system increases as the gravitational potential energy decreases, with the total mechanical energy remaining constant (neglecting air resistance). However, when the book hits the ground, it comes to rest and mechanical energy is "lost." Energy just can't disappear, so where did it go? Some goes to sound energy as an audible "thud"

https://doi.org/10.1515/9783110642292-007

is created; however, most of the mechanical energy is transferred to internal energy. More precisely, the collision jostles the molecules of the book and ground, increasing their random, microscopic motion. This internal energy is really a thermal energy, and as it increases, the temperature of the book and ground also increase, even if just slightly.

The human body also possesses internal energy, but it is more complex than the random thermal motion of the molecules in a lifeless book. Sure, some of the body's internal energy is random thermal motion, but other more organized energy transfers occur to sustain life processes like breathing and circulating blood as well as to activate skeletal muscles. Ultimately this energy comes from energy stored in the chemical bonds of the food we eat. *Metabolic energy* is the chemical energy used by the body to perform all its functions. Metabolic energy cannot be ignored when the dancer wonders about the overall energy requirements for a game of tennis and her question about why she must breathe so hard during vigorous exercise.

This chapter covers both mechanical and metabolic energy considerations in dance, beginning with work and mechanical energy for different choices of systems (e. g., single dancer, a dancer and the Earth, an individual body segment, etc.) and how different systems can provide us with different information or ways to solve problems. Metabolic energy will then be covered, including factors that contribute to the rate at which it is used, how metabolic energy usage is measured experimentally, and factors that contribute to efficient (or inefficient) metabolic energy use.

7.1 Mechanical work

7.1.1 Mechanical work done by a force

Let's begin by choosing our system of interest to be a single dancer and use our simplest physical model: a point particle at the CM. An external force $\vec{F}$ acting on a particle as it moves an infinitesimal displacement $d\vec{x}$ does an infinitesimal amount of work, defined by the scalar product $dW = \vec{F} \cdot d\vec{x}$, since it is the component of $\vec{F}$ parallel to $d\vec{x}$ that contributes to the work done.

$$dW = F \cos \phi \, dx = F_{\parallel} \, dx$$

where ϕ is the angle between $\vec{F}$ and $d\vec{x}$. To find the total *work* done by the force over a finite displacement from point A to point B, we must add up (i. e., integrate) all of the infinitesimal bits of work along the path.

$$W = \int_{A}^{B} \vec{F} \cdot d\vec{x} \tag{7.1}$$

This is a line integral which depends on the path taken from points A to B, and the force can in general be non-constant (e. g., as is most often the case for the GRF). In Cartesian coordinates eq. (7.1) can be written as $W = \int F_x\, dx + \int F_y\, dy + \int F_z\, dz$. For 1-D motion (e. g., vertical, or z-direction motion), the problem is simplified to $W = \int F_z\, dz$ and further simplified if the force F_z is constant, $W = F_z\Delta z = F_z(z_B - z_A)$. Although both $\vec{F}$ and $d\vec{x}$ are vectors, their dot product, work, is a scalar with units of Joules (1 N m = 1 J). Work done by a force can be positive or negative depending on whether the force has a component parallel or anti-parallel to the displacement. The positive or negative sign does not indicate a direction (work is a scalar, after all!), but as we will find in Section 7.2, forces that do negative work tend to decrease the kinetic energy and forces that do positive work tend to increase the kinetic energy of the system.

To find the total (or net) work done by all forces on a dancer, eq. (7.1) becomes

$$\Sigma W = \int_A^B \Sigma\vec{F} \cdot d\vec{x}$$

In solo dancing without props and considering the dancer as a point particle at the CM, only the external forces of gravity and the GRF (normal force (F_N) and force of friction (F_f)) may do work on the dancer (neglecting any very small work done by air resistance). Therefore, $\Sigma W = \int F_{f,x}\, dx + \int F_{f,y}\, dy + \int (F_N - Mg)\, dz$ for horizontal ground. If the motion of the dancer's CM is purely vertical, only F_N and F_g can do work. If the motion of the dancer's CM is purely horizontal, only F_f can do work. Not often is CM movement purely in a single direction in dance and human movement more generally, but there are cases in which we wish to only analyze a single direction.

Example. A dancer (m = 61.1 kg) is pushing off for a vertical leap as shown in Fig. 7.1. The dancer does a countermovement jump: beginning with the legs straight at point A (z_A = 0.95 m), bending the knees for a downward motion of the CM to point B (z_B = 0.73 m), and then immediately straightening the legs to move the CM upward to point C. (Assume for now that $z_C = z_A$ = 0.95 m.) Find the work done by gravity on the dancer from A to B, from B to C, and the total work done by gravity for the entire pushoff (A to B to C).

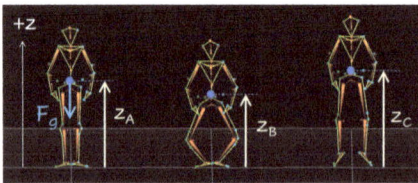

Figure 7.1: Center of mass vertical positions (z) during pushoff for a vertical jump. The constant force of gravity vector acts vertically downward (−z-direction).

Solution. From A to B, the force of gravity and the CM displacement are both in the negative z-direction, so the work done by gravity on the dancer is positive.

$$W_{g,A \to B} = \int_A^B \vec{F}_g \cdot d\vec{x} = \int_{z_A}^{z_B} -Mg \, dz = -Mg(z_B - z_A)$$

$$= -(61.1 \, \text{kg})(9.81 \, \text{m/s}^2)(0.73 \, \text{m} - 0.95 \, \text{m})$$

$$= 131.9 \, \text{J}$$

From B to C, the force of gravity remains in the negative z-direction, but the displacement is in the positive z-direction, so the work done by gravity is negative.

$$W_{g,B \to C} = \int_B^C \vec{F}_g \cdot d\vec{x} = -Mg(z_C - z_B)$$

$$= -(61.1 \, \text{kg})(9.81 \, \text{m/s}^2)(0.95 \, \text{m} - 0.73 \, \text{m})$$

$$= -131.9 \, \text{J}$$

To find the total work done by gravity over the entire path of the CM, we simply sum the work done over the individual parts.

$$W_{g,A \to C} = W_{g,A \to B} + W_{g,B \to C} = 131.9 \, \text{J} + (-131.9 \, \text{J}) = 0 \, \text{J}$$

The total work by gravity over this closed path (starting and ending at the same point) is zero. This result for gravity is not specific to the case presented here. No matter the path taken by the dancer, if the CM begins and ends in the same vertical position, the total work done by gravity will be zero. Forces such as this are termed "conservative" and the implications will be discussed in more detail in Section 7.4. (Note that z_C may be slightly greater than z_A since the dancers ankles plantarflex before her feet completely leave the ground. This additional negative work done by gravity can be accounted for in the calculation.)

Example. The vertical ground reaction force (vGRF) was measured with a force plate during another countermovement jump performed by the dancer in the previous example (Fig. 7.1). The vertical CM position was simultaneously tracked during the movement. Use the data collected to find the work done by vGRF on the dancer from A to B, from B to C, and the total work done by vGRF for the entire pushoff (A to B to C).

Solution. First thinking conceptually, we know that from A to B the work done by vGRF must be negative, since vGRF is vertically up and the displacement is vertically down. We also then know that from B to C the work done by vGRF must be positive, since the force and displacement are in the same direction. This is the opposite of the sign of the work done by gravity during these phases of the pushoff.

We also notice that this problem is more complicated than the previous example since vGRF is non-constant. So we must evaluate the work done via the integral in eq. (7.1), or $W_{vGRF} = \int (vGRF)\, dz$. A plot of vGRF vs. z (Fig. 7.2) allows us to evaluate the integral above by finding the area under the curve. (We used a similar process in Chapter 3 to find the impulse from the area under a force vs. time curve.) The overall motion begins and ends at nearly the same vertical position (A and C), so arrows are drawn on the plot to show how the force changes from A to B to C. The (negative) work done from A to B ($W_{A \to B}$) is shown with the orange shading and the (positive) work done from B to C ($W_{B \to C}$) is shown with the blue shading (from the upper curve all the way down to the z-axis). The total work done is then $W_{vGRF} = W_{A \to B} + W_{B \to C}$ (which will be approximately equal to the area in the region between the two curves).

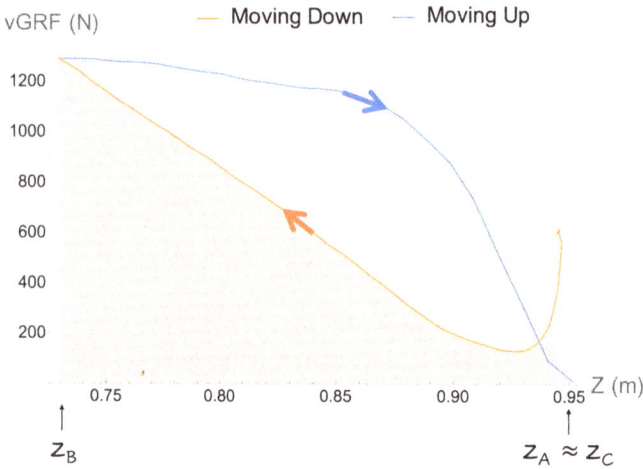

Figure 7.2: Vertical ground reaction force (vGRF) as a function of CM position (z) during pushoff for a vertical jump. The orange curve is vGRF when the dancer is moving from A to B (CM going down) and the blue curve is vGRF when the dancer is moving from B to C (CM going up).

Recall, our data are discrete (Tab. 7.1), so to find the area under the curve in practice, we will sum all of the small amounts of work done (ΔW) over small displacements (Δz). This time each Δz is not the same (Δt is set by the capture rate of the data, but how much the CM moves vertically in this fixed amount of time changes depending on the dancer's velocity).

We then compute the small amount of work ΔW during each Δt between adjacent data points by calculating

$$\Delta W = \bar{F}_z \Delta z \tag{7.2}$$

where $\Delta z = z_{i+1} - z_i$ and $\bar{F}_z$ is the mean vGRF from t_i to t_{i+1} (Tab. 7.2). Note that this method can be utilized even if the sampling frequencies (capture rate) of the GRF data

Table 7.1: Sample from vertical CM position (z) and vertical ground reaction force (vGRF) data measured during pushoff for a vertical jump. The entire pushoff occurred from $t = 0$ to 1.056 s. Times t_{144} to t_{148} are within the range of time that z is moving vertically down, and times t_{190} to t_{194} are within the range that z is moving vertically up.

t_i	t (s)	z (m)	vGRF (N)
⋮	⋮	⋮	⋮
t_{144}	1.046	0.7833	520.1
t_{145}	1.052	0.7719	557.8
t_{146}	1.059	0.7601	605.3
t_{147}	1.066	0.7483	657.4
t_{148}	1.072	0.7363	694.8
⋮	⋮	⋮	⋮
t_{190}	1.352	0.6398	1270.9
t_{191}	1.359	0.6470	1263.6
t_{192}	1.366	0.6544	1256.3
t_{193}	1.372	0.6624	1251.1
t_{194}	1.379	0.6709	1243.9
⋮	⋮	⋮	⋮

Table 7.2: Vertical displacement Δz and mean vGRF ($\bar{F}_z$) during individual time intervals computed from the data in Tab. 7.1. The amount of work ΔW during each time interval is also computed. ΔW are negative for the times that the CM moves vertically down and positive when the CM moves up as expected.

Δz (m)	$\bar{F}_z$ (N)	ΔW (J)
⋮	⋮	⋮
−0.01139	539.0	−6.139
−0.01174	581.6	−6.828
−0.01186	631.4	−7.485
−0.01194	676.1	−8.076
⋮	⋮	⋮
0.0071	1267.2	9.054
0.0074	1259.9	9.371
0.0080	1253.7	9.973
0.0085	1247.5	10.62
⋮	⋮	⋮

and motion capture data are not the same, as long as the data are triggered to be captured simultaneously with a common time origin. We would simply find the mean vGRF over the same time interval as the corresponding Δz.

Then, the total work during the entire pushoff is found by summing all of the work done during each time interval.

$$W = \Sigma \Delta W$$

Numerically, this value was computed to be $W = 157.4$ J. (Note: The sum of all of the ΔW values while the CM was moving down was -246.9 J and when the CM was moving up it was 404.3 J for an overall total of 157.4 J.) Unlike the work done by gravity, the total work done by vGRF is positive during the pushoff for the jump, as we predicted conceptually.

The analysis applied in the previous examples to find the work done by external forces on the dancer's CM can be extended to analyze 2-D or 3-D motion, for example, to find the work done by the total GRF vector during the takeoff for a horizontal leap. The difference would be that to find the ΔW during each discrete time interval, eq. (7.2) would become

$$\Delta W = \bar{F}_x \Delta x + \bar{F}_y \Delta y + \bar{F}_z \Delta z$$

using scalar addition, where F_x and F_y are the two components of the horizontal GRF, and Δx and Δy are the displacements of the CM in those directions.

7.1.2 Mechanical work done by a torque

Now we will shift our focus to a different system: an individual body segment. For example, if a dancer does a straight leg kick with no relative motion between the thigh, shank, and foot, we can model the entire leg segment as a rigid body rotating around an axis at the hip joint. Now our system is no longer a point particle with translational motion, but a rigid body with rotational motion around the joint axis (assuming the hip joint remains stationary). Similar to the work done by a force over a linear displacement, the work done by a torque over an angular displacement is

$$W = \int_A^B \vec{\tau} \cdot d\vec{\theta} \tag{7.3}$$

If the torque is constant, the work done is $W = \vec{\tau} \cdot \Delta\vec{\theta} = \tau\Delta\theta \cos\phi$ where ϕ is the angle between the torque and the angular displacement. If the torque and angular displacement have components in the same direction, the work done by the torque is positive ($\phi = 0$, so $\cos\phi = 1$). Conversely, if the torque and angular displacement have components in opposite directions, then the work done by the torque is negative ($\phi = \pi$, so $\cos\phi = -1$).

Let's imagine a dancer doing a front kick in the sagittal plane, where we define the flexion direction as positive and extension as negative (Fig. 7.3). During hip flexion, the hip flexors provide a torque in the positive direction, so they do positive work on the leg. But as the hip extensors activate and exert a torque in the negative direction ($+\Delta\theta$ and $-\tau$), the hip extensors do negative work on the leg. In general, concentric activation contributes to positive work and eccentric activation contributes to negative work done on a segment. During isometric activation, there is no relative motion between segments ($\Delta\theta = 0$). This means no mechanical work is done by muscles activated isometrically.

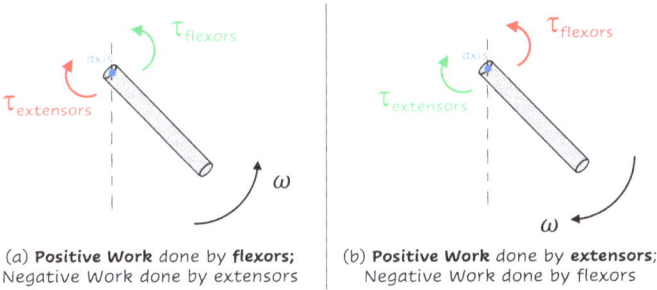

(a) **Positive Work** done by **flexors**; Negative Work done by extensors

(b) **Positive Work** done by **extensors**; Negative Work done by flexors

Figure 7.3: Gesture leg segment during a front kick (sagittal plane view). The supporting leg (not shown) would be aligned with the dashed vertical line, and the hip joint center (axis of rotation) is shown as a blue point. When the leg's angular velocity (ω) is counterclockwise (a), the hip flexors do positive work, whereas the hip extensors do negative work on the leg since their torque is in the direction opposite of the angular displacement. The reverse is true for hip extension (ω clockwise) as shown in (b).

Example. Find the work done by gravity on the leg during a front kick that moves from $\theta = 0°$ of hip flexion to $135°$ of hip flexion. Assume the same dancer performs this kick as done in Chapter 6, so the mass of the dancer's leg and its CM position from the hip are known ($m = 12.92$ kg, $r = 0.278$ m).

Solution. With the hip flexion direction defined as positive, the negative torque due to gravity is

$$\vec{\tau}_g = \vec{r} \times \vec{F}_g$$
$$\tau_g = -rmg \sin\theta$$

where m is the mass of the leg segment, r is the radial position of the leg's CM from the hip joint axis of rotation and θ is the hip flexion angle. The torque due to gravity on the leg increases as hip flexion moves from $0°$ to $90°$, where the moment arm is greatest. Then it begins to decrease again when the hip flexion angle (θ) moves from $90°$ to $135°$ (Fig. 7.4).

To find the work done by gravity during the entire kick, we will utilize eq. (7.3).

$$W_g = \int_{\theta_i}^{\theta_f} \vec{\tau}_g \cdot d\vec{\theta}$$

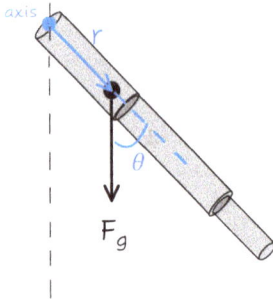

Figure 7.4: Gesture leg segment during a front kick (sagittal plane view), with the force of gravity acting vertically downward from the leg's CM position. Hip joint center (axis of rotation) is shown as a blue point.

$$= \int_{\theta_i}^{\theta_f} -rmg \sin\theta \, d\theta = -rmg \int_{\theta_i}^{\theta_f} \sin\theta \, d\theta$$

$$= -rmg[-\cos\theta]\big|_{\theta_i}^{\theta_f}$$

$$W_g = rmg(\cos\theta_f - \cos\theta_i)$$

Substituting in numerical values for r, m, g, initial and final hip flexion angles, we have,

$$W_g = (0.278\,\text{m})(12.92\,\text{kg})(9.81\,\text{m/s}^2)(\cos 135° - \cos 0°)$$

$$= -60.1\,\text{J}$$

We solved for the work done by gravity on the leg, but what about the work done by the muscle torques? Given measurements of the leg's inertia and rotational acceleration, we could use our inverse dynamics approach to find the net work done by the muscles. Since $\Sigma\vec{\tau} = \Sigma\vec{\tau}_m + \vec{\tau}_g$, we know that the net muscle torque is $\Sigma\vec{\tau}_m = \Sigma\vec{\tau} - \vec{\tau}_g$. Substituting $I\vec{\alpha}$ for the net torque $\Sigma\vec{\tau}$, we have

$$\Sigma\vec{\tau}_m = I\vec{\alpha} - \vec{\tau}_g$$

and using the definition of work in eq. (7.3)

$$\Sigma W_m = \int_A^B \Sigma\vec{\tau}_m \cdot d\vec{\theta}$$

$$= \int_A^B I\vec{\alpha} \cdot d\vec{\theta} - \int_A^B \Sigma\vec{\tau}_g \cdot d\vec{\theta}$$

$$= \int_A^B I\vec{\alpha} \cdot d\vec{\theta} - W_g$$

If inverse dynamics is combined with a musculoskeletal model to estimate individual muscle forces as demonstrated in Sections 6.4 and 6.5, we can then compute work done by individual muscles. For example, given graphical results of iliopsoas muscle torque versus hip flexion angle, we could find work done by the iliopsoas with the area under the curve in Fig. 6.13.

We discussed how to find the work done on a system, but what sort of useful information can work provide? To find out, we once again begin with Newton's second law and develop the work-kinetic energy theorem.

7.2 Work-kinetic energy theorem

7.2.1 Translational motion

Similar to rewriting Newton's second law into the impulse-momentum theorem in Chapter 3, we can repackage Newton's second law into the work-kinetic energy theorem. This will enable us once again to link a cause with an effect. Recall Newton's second law can be read as stating "A net external force causes a mass to accelerate," and the impulse-momentum theorem states that "A net impulse causes a system to change momentum." We will soon see that net work done on a system causes a change in its kinetic energy.

Here we begin with Newton's second law in the form $\Sigma \vec{F} = m\vec{a}$ since mass is constant for all of our biomechanical systems, and assume for now that the motion is in 1-D (e. g., the x-direction).

$$\Sigma F = m\frac{dv}{dt}$$

If we multiply the right hand side by $1 = \frac{dx}{dx}$, perform some algebra, and use the definition of velocity ($v = \frac{dx}{dt}$),

$$\Sigma F = m\frac{dv}{dt}\frac{dx}{dx} = m\frac{dv}{dx}\frac{dx}{dt}$$

$$\rightarrow \Sigma F = mv\frac{dv}{dx}$$

Then, we multiply both sides by dx and integrate.

$$(\Sigma F)\,dx = mv\,dv$$

$$\int_{x_i}^{x_f} (\Sigma F)\,dx = \int_{v_i}^{v_f} mv\,dv$$

$$\int_{x_i}^{x_f} (\Sigma F)\,dx = \frac{1}{2}mv_f^2 - \frac{1}{2}mv_i^2$$

The quantity on the left hand side is the net work done on the system by the sum of all forces, and since kinetic energy is defined as $K = \frac{1}{2}mv^2$, the quantity on the right hand side is the change in kinetic energy $\Delta K = K_f - K_i$. The result generalizes to 3-D motion, in which we use the scalar product to find net work done, so the *work-kinetic energy theorem* becomes:

$$\int_{\vec{x}_i}^{\vec{x}_f} (\Sigma \vec{F}) \cdot d\vec{x} = \frac{1}{2}mv_f^2 - \frac{1}{2}mv_i^2 \tag{7.4}$$

$$\Sigma W = \Delta K$$

From eq. (7.4) we see that if net work done on a system is negative, the kinetic energy of the system will decrease (energy is taken out of the system). If net work is positive, then kinetic energy increases (energy is added into the system).

Example. Use the results found in Section 7.1.1 for the work done by gravity and vGRF during vertical jump pushoff to find the takeoff velocity of the dancer as her feet leave the ground (assume she starts from rest at point A). Then find the vertical displacement of the CM from the point at which the feet leave the ground to the peak of the jump.

Solution. Since we already found the work done by the two external forces on the dancer, we simply add them up and set the total work equal to the dancer's change in kinetic energy.

$$\sum W = \Delta K$$

$$W_g + W_{\text{vGRF}} = \frac{1}{2}mv_f^2 - 0$$

$$0\,\text{J} + 157.4\,\text{J} = \frac{1}{2}(61.1\,\text{kg})v_f^2$$

$$\rightarrow v_f = 2.27\,\text{m/s}$$

This is the vertical speed of the dancer's CM the moment her feet leave the ground.

Once her feet leave the ground, her CM acceleration is $g = -9.81\,\text{m/s}^2$ (neglecting air resistance), and kinematics can be used to determine the vertical CM displacement to the peak of the jump (where $v = 0$). For the "free flight" phase of the jump, the take-off velocity is $v_0 = 2.27\,\text{m/s}$.

$$v^2 = v_0^2 + 2a\Delta z$$

$$0^2 = (2.27\,\text{m/s})^2 + 2(-9.81\,\text{m/s}^2)\Delta z$$

$$\rightarrow \Delta z = 0.26\,\text{m}$$

It is worth noting that the actual vertical CM position when this dancer's feet left the ground (when vGRF $\longrightarrow$ 0) was slightly higher than her CM position at the beginning of the pushoff phase ($z_C > z_A$). Using the precise values, the true work done by gravity $W_g = -7.85\,\text{J}$. If we correct for this in our above analysis, the true takeoff velocity is computed to be 2.21 m/s, only 2.5 % less than the value computed assuming $z_A = z_C$. This small difference is important, however, for precise calculations when conducting research.

7.2.2 Rotational motion

For pure rotational motion, net work done still leads to a change in kinetic energy. More specifically, the net work done by a torque leads to a change in *rotational* kinetic energy.

$$\sum W = \Delta K$$

$$\int_{\vec{\theta}_i}^{\vec{\theta}_f} (\Sigma \vec{\tau}) \cdot \mathrm{d}\vec{\theta} = \frac{1}{2} I \omega_f^2 - \frac{1}{2} I \omega_i^2 \tag{7.5}$$

In this case, I is the moment of inertia and ω is the rotational speed of the rigid body. Again, this version of the work-kinetic energy theorem is particularly useful when dealing with torques on body segments around joint axes of rotation.

Example. Find the work done by the net muscle torque around the hip joint during the front kick in the example in Section 7.1.2 (a) during hip flexion (from 0° to 135°), and (b) during hip extension (from 135° back down to 0°). Assume that the entire front kick (flexion followed by extension) begins and ends with the dancer's gesture leg at rest at 0° (vertically down).

Solution. During hip flexion (a) the dancer's leg begins and ends at rest ($\omega_i = \omega_f = 0$ rad/s). (When the full 135° of hip flexion is reached, the dancer's leg comes momentarily to rest as direction changes from flexion to extension.) Once we realize this fact, computing the work done by the net muscle torque becomes a straightforward application of eq. (7.5) with $\Delta K = 0$.

$$\sum W = \Delta K$$
$$W_g + \Sigma W_m = 0$$

This means that the work done by the net muscle torque (ΣW_m) is the negative of the work done by the torque due to gravity on the leg. As we found in Section 7.1.2, gravity does negative work when the leg moves upward during flexion. So, the net muscle torque does positive work on the leg segment during hip flexion.

$$\Sigma W_m = -W_g$$
$$= -(-60.1\,\mathrm{J})$$
$$= 60.1\,\mathrm{J} \quad \text{(flexion)}$$

When the leg moves back toward the ground during hip extension (b), it begins and ends at rest once again ($\Delta K = 0$), so $\Sigma W_m = -W_g$ as before. During hip extension, however, the work done by the gravitational torque is positive, so the net muscle work is negative.

$$\Sigma W_m = -60.1\,\text{J} \quad \text{(extension)}$$

Notice that this result only depended upon the initial and final angular positions, but not *how* the straight-legged kick was performed. The same net muscle work is required whether the kick is performed in parallel or turnout or whether at a high velocity or slow and controlled. The important point to emphasize is that this result is because we have calculated the work done by all of the muscles spanning the hip joint. Which muscles contribute to flexion or extension and the amount that they are activated depends on the amount of hip external rotation (turnout) among other factors. We would encounter the same issues as Chapter 6 (i. e., problem of indeterminacy) if we attempt to determine rotational work done by an individual muscle spanning a joint. Instead, we will get a better conceptual sense for which muscles perform work and when in the following conceptual example.

Example. Discuss qualitatively how the muscle groups (hip flexors and hip extensors) do rotational work on the leg segment during a front kick performed under two different conditions (1) rapid flexion followed by a slower, more controlled extension, and (2) slow and controlled for both flexion and extension. Fig. 7.5 illustrates the different conditions.

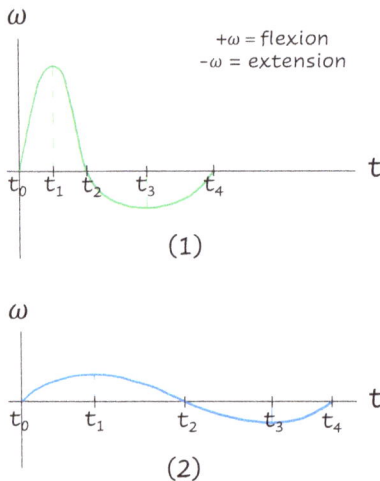

Figure 7.5: Rotational velocity (ω) vs. time for a front kick performed with (1) rapid flexion, controlled extension, and (2) slow and controlled flexion and extension. The axes of the two graphs are on the same scale for comparison. At t_0 the leg is at 0° flexion, reaching maximum velocity at t_1. The full 135° flexion is reached at t_2, where the rotational velocity goes through zero. From t_2 to t_4 the leg is being lowered (hip extension back to 0° at t_4 and the maximum extension velocity occurs at t_3).

Solution. From t_0 to t_1 the leg moves from rest to its maximum upward velocity. (With condition (1) being a quick kick, the peak at t_1 is large.) The flexors do positive work (W_f) and the extensors, if coactivated, would do negative work (W_e) along with the negative work done by gravity (W_g). We know the net work ($\Sigma W = W_f + W_e + W_g$) must be positive since the kinetic energy is increasing from t_0 to t_1, so $|W_f| > |W_e + W_g|$. For condition (2), maximum ω is lower so the flexors must do less work to increase the leg's kinetic energy than condition (1) (assuming W_e is small for each case).

From t_1 to t_2, the leg is still moving upward ($\omega > 0$), so *if* the flexors and extensors are coactivating, $W_f > 0$ and $W_e < 0$ as before. However, the leg's rotational speed is decreasing as it approaches its peak angular displacement (θ_{max}), which means $\Delta K < 0$, and therefore the negative work done by gravity and active extensors must exceed the positive work done by any flexors if activated. From t_1 to t_2 the flexors may have a lesser activation during a rapid kick so that the large negative angular acceleration needed to stop the leg at the top can be achieved. On the other hand, for the slow leg lift in condition (2) the flexors remain active but such that $|W_e + W_g|$ is only slightly greater than $|W_f|$. If θ_{max} is not approaching the dancer's limit in hip flexion range of motion, and the large negative angular acceleration in condition (1) cannot be achieved with the torque due to gravity alone, then the extensors must be eccentrically activated to do the negative work required. If however θ_{max} does approach the dancer's limit in hip flexion range of motion, then capsuloligamentous forces surrounding the joint structure can provide negative torque with less involvement from the extensors.

When the rotational velocity changes to moving in the negative (extension) direction, any activated extensors do so concentrically and now do positive work along with gravity. From t_2 to t_3, K increases so $|W_e + W_g| > |W_f|$, but from t_3 to t_4 when K decreases, the negative work done by the flexors must exceed the work done by the extensors and gravity to stop the leg's motion and allow the foot to gently come to rest on the ground. If the extensors were activated greatly during the entire time of hip extension, the leg would continue to have a large negative acceleration, speeding up during the entire hip extension, and slam into the ground at a high velocity.

No matter how the kick is executed, since $\sum W_g = 0$ from start to finish, the net mechanical work done by the muscles is also zero. Different kicking conditions require different muscle activation, however, with different metabolic energy requirements. During the front kick, muscles other than flexors and extensors (the agonists/antagonists) also exert force. These include fixator muscles that are activated to *prevent* unwanted rotation of segments. For example, the abdominal muscles must be engaged to prevent pelvic tilting when the hip flexors are active (Fig. 7.6). Since the pelvis and trunk segments remain at rest, the abdominal muscles do no work. Metabolic energy is required to activate fixator muscles such as this, even if they do zero mechanical work. Similarly, statically held poses require zero mechanical work done by muscles, although there is nonzero metabolic energy required to activate the muscles.

A

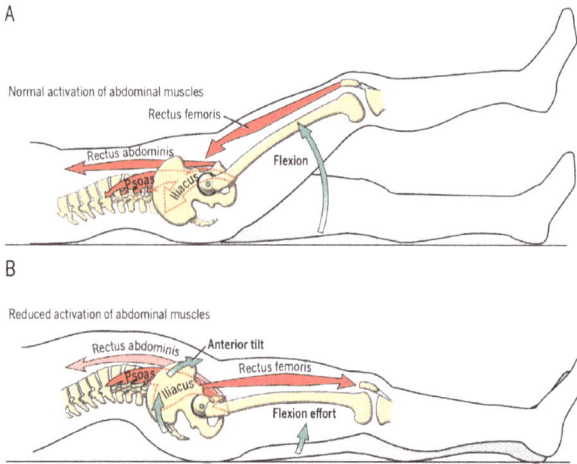

B

Figure 7.6: Activated abdominal muscles fixate the pelvis segment so that it does not rotate when the hip flexors are activated (top figure). If the abdominals are not sufficiently activated, the hip flexors rotate the pelvis anteriorly (bottom figure), increasing sway in the back (lumbar lordosis). Reprinted with permission from [73].

7.3 Power

A qualitative descriptor such as "powerful" may be used to describe a dance movement like an explosive jump. One dancer's leap might appear more powerful than another's, and there is likely a measurable difference in the biomechanical power generated between the two dancers' leaps to affirm the qualitative difference. The definition of *power* is the rate at which work is done, with units of watts (1 J/s = 1 W).

$$P = \frac{dW}{dt} \tag{7.6}$$

Eq. (7.6) is the instantaneous power, whereas the average power would be computed from the total work and the time to do that work ($P_{avg} = \frac{W}{t}$).

An alternate definition of power is the rate at which energy is transferred ($P = \frac{dE}{dt}$) into or out of a system or between different forms of energy. For example, the power rating of a light bulb is how much electrical energy is transferred to light and heat energy per second. A 100-W light bulb uses 100 J of electrical energy every second, and a 60-W bulb uses 60 J per second. Whereas a homeowner may wish to reduce energy consumption by using a lower wattage light bulb, a dancer may wish to increase the power of a jump so it will appear more impressive. How might we quantify the power rating of something like a jump, however?

One difference (of many!) between the light bulb and a dancer is that the rate at which energy is used by the light bulb is approximately constant. This is most often not true for a dancer. The power in eq. (7.6) is really a function of time, $P(t)$. Another way to compute instantaneous power is by considering the infinitesimal bit of work done by a force over a small displacement, $dW = \vec{F} \cdot d\vec{x}$. Then the instantaneous power is

$$P = \frac{dW}{dt} = \vec{F} \cdot \frac{d\vec{x}}{dt} = \vec{F} \cdot \vec{v} \tag{7.7}$$

Power, like work, is computed with the dot product, $P = \vec{F} \cdot \vec{v} = Fv \cos \theta = F_x v_x + F_y v_y + F_z v_z$.

Eq. (7.7) can be used to find the power *generated* or *absorbed* by external forces acting on a dancer's center of mass system. If the force acts in the same direction as the dancer's CM velocity (or has a component in the same direction), $P > 0$ and the force generates power. The ground reaction force generates power during the pushoff for a jump, doing work to increase the dancer's kinetic energy at take off. A partner's lifting force generates power to raise a dancer's CM high off of the ground. If the force acts in the direction opposite of the CM velocity (or has a component in the opposite direction), $P < 0$ and the force absorbs power. The GRF absorbs power during the landing from a jump. A partner can absorb power as they catch a dancer or help to cushion their landing from a lift.

Example. Compute the instantaneous power generated and absorbed by the vGRF during (a) the push off and (b) landing of a jump. Plot power vs. time, find the maximum power, and average power for the two situations. Find the total mechanical energy generated from the vGRF during push off.

Solution. From digitized data of vGRF and z_{CM} (like that in Tab. 7.1), first the vertical velocity ($\dot{z}$) is computed using finite difference method (eq. (3.6)), then digitized power is computed from eq. (7.7), $P_i = F_i \dot{z}_i$. Results are displayed in Fig. 7.7.

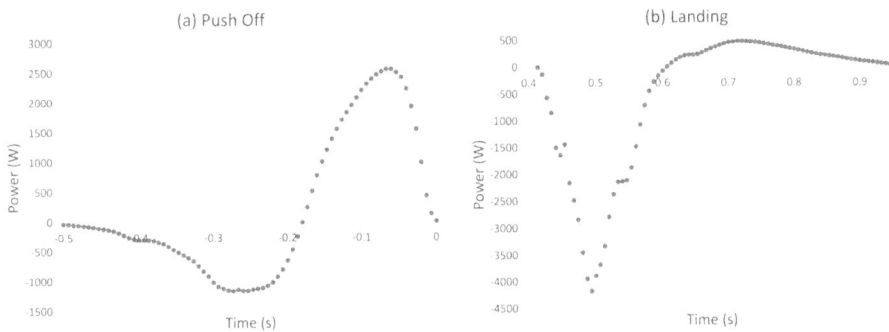

Figure 7.7: vGRF power vs. time during (a) pushoff and (b) landing of a countermovement jump. Time $t = 0$ was defined as the time at which the dancer's feet leave the ground, so negative time denotes push off for the jump. The dancer was in the air for 0.4 seconds (the first data point in (b) is when the dancer's feet make contact with the ground for landing after the jump).

Since this was a countermovement jump, during push off (a), vGRF first absorbs power ($P < 0$) as the CM moves vertically downward, then power is generated ($P > 0$) by the vGRF with $P_{max} \approx 2600$ W. The average generated power during pushoff (during the time when $P > 0$) was $P_{avg} \approx 1600$ W. During landing (b), vGRF absorbs power with a peak value of $P_{max} \approx -4200$ W (the most negative value is the maximum power absorbed), and the average power absorbed (when $P < 0$) was $P_{avg} \approx -1800$ W.

Since $P = \frac{dE}{dt}$, the energy generated by vGRF during pushoff can be computed from $E = \int P\,dt$, or the area under the power vs. time curve. Numerically, the area under the negative portion of the power vs. time curve was -180 J and the area under the positive portion was 290 J resulting in a total of 110 J of energy generated by vGRF during takeoff.

7.3.1 Net joint power

As we learned previously, a torque can do work through an angular displacement, so we can also determine the power or rate at which work is done during rotational motion. Starting with the definition of power in eq. (7.6), and noting in this case that $dW = \vec{\tau} \cdot d\vec{\theta}$,

$$P = \frac{dW}{dt} = \vec{\tau} \cdot \frac{d\vec{\theta}}{dt} = \vec{\tau} \cdot \vec{\omega} \tag{7.8}$$

This version of power is often useful in biomechanics to find the rate at which energy is generated or absorbed at joints. The total power at a joint is found from the net torque exerted around that joint.

$$\Sigma P_{joint} = \Sigma \vec{\tau}_{joint} \cdot \vec{\omega} \tag{7.9}$$

Since the net torque is proportional to the joint angular acceleration, the power profile (normalized by inertia, or kg·m^2) can be computed from

$$\Sigma P_{joint} = I\vec{\alpha} \cdot \vec{\omega}$$
$$\Sigma P_{joint} \propto \vec{\alpha} \cdot \vec{\omega} \tag{7.10}$$

The net joint power includes flows of energy to and from the segments spanning the joint due all forces that produce torque. This can include muscle torques, torques due to gravity, or other external torques (e. g., ground reaction force acting on the foot segment produces a torque around the ankle joint, etc.).

Example. From experimental measures of joint rotational velocity and acceleration, determine the net joint power at the ankle, knee, and hip during a dancer's landing from a jump.

Solution. For this example, we will focus our attention on the flexion/extension kinematics of the jump landing. The normalized net power was computed from eq. (7.10) for each joint. Results are displayed in Fig. 7.8. With ΣP_{ankle}, ΣP_{knee}, and ΣP_{hip} plotted on the same axes, we first observe that the maximum rate of energy transfer into ($P > 0$) and out of ($P < 0$) the segments spanning the ankle is greater than that for the knee or hip. $\Sigma P_{ankle,max} \approx 2.5(\Sigma P_{knee,max})$ with ΣP_{hip} very small compared to the other net joint powers. It makes sense that $\Sigma P_{ankle} > \Sigma P_{knee} > \Sigma P_{hip}$ since the ankle is closest to the ground. The foot segment absorbs the most energy from the floor due to impact, and the lower leg and thigh absorb less energy since they are further from the contact point with the ground.

ΣP_{ankle} and ΣP_{knee} follow roughly the same net joint power profile. Each joint power is initially positive as both the joint rotational velocities and accelerations are in the same direction. When the foot initially makes contact with the ground during landing, the ankle dorsiflexes with increasing rotational velocity in the dorsiflexion direction. The knees begin to flex, also with increasing flexion velocity. Then, as the activated muscles slow the ankle dorsiflexion and knee flexion (beginning at $t \approx 0.13$ s), the kinetic energy of the segments spanning these joints decrease and there is net joint power output (negative) from the ankle and knee. Once the maximum plié is reached (at $t \approx 0.30$ s), knee extension and ankle plantarflexion commence, and some kinetic energy is transferred back into the lower limbs; however, the rate at which energy transfer occurs is now much less that that during initial impact, so the net joint powers are much lower when $t > 0.3$ s.

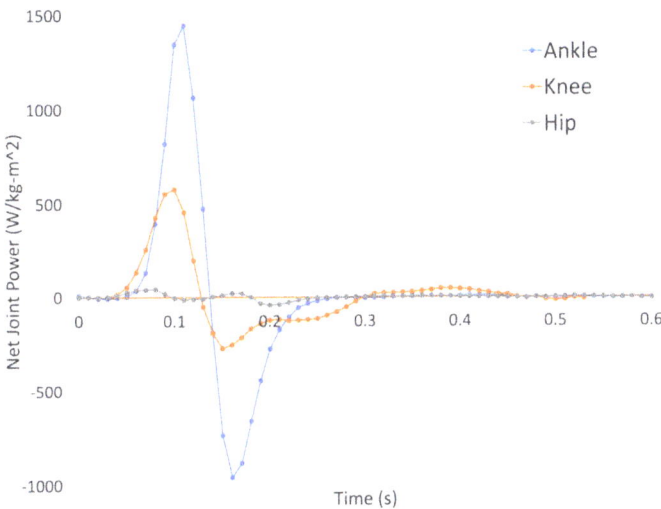

Figure 7.8: Net joint power at the ankle, knee, and hip during a dancer's landing from a vertical jump.

7.3.2 Net muscle power at a joint

In the previous example, the ground reaction force exerts a large torque around the ankle joint during a jump's landing, so it has a large contribution to the overall energy input to the foot segment system. This results in a positive power input to the ankle joint from the GRF. We can instead isolate the *muscle* power at a joint to determine the rate at which the muscles themselves generate or absorb energy at a joint. The net torque on a body segment consists of the net muscle torque and the torques due to outside forces. Keeping with the example of the net torque at the ankle joint during the landing from a jump, the net muscle torque can be computed from the following.

$$\Sigma \vec{\tau} = I\vec{a}$$
$$\vec{\tau}_{\text{GRF}} + \vec{\tau}_g + \Sigma \vec{\tau}_m = I\vec{a}$$
$$\Sigma \vec{\tau}_m = I\vec{a} - \vec{\tau}_{\text{GRF}} - \vec{\tau}_g$$

The torque due to the GRF is found from $\vec{\tau}_{\text{GRF}} = \vec{r}_{\text{GRF}} \times \overrightarrow{GRF}$, where $\vec{r}_{\text{GRF}}$ is the vector position from the ankle joint center to the center of pressure (CP) position on the foot. The CP can be measured with a force plate beneath the foot. (Note that we would need a single force plate beneath each foot to isolate the GRF vector on a single foot segment.) The (relatively small) torque due to the weight of the foot segment is found from $\vec{\tau}_g = \vec{r}_g \times m_f\vec{g}$, where $\vec{r}_g$ is the vector position from the ankle joint center to the center of mass of the foot segment.

During the initial impact with the ground, the ankle plantarflexors are active, producing a counterclockwise torque in Fig. 7.9 to oppose the large clockwise torque due to the GRF. Thus, the net *muscle* power around the ankle is negative, absorbing energy during the impact of landing from a jump. One study found the net muscle power around the ankle joint during landings from jumps on 5 different flooring types (Fig. 7.10) [35]. The peak muscle power at the ankle was the least for the floor with the greatest vertical deformation during landings. The floor helped absorb some of the energy that the foot and ankle otherwise would have.

Another study compared the net muscle power at the ankle joint during pushoff for two different types of ballet leaps, the saut de chat (Fig. 7.11) and the temps levé [77]. The temps levé is also a leap from one foot to one foot, however, the leap is vertical instead of a 2D leap like the saut de chat. A significant difference ($p < 0.001$) between the two types of leaps in peak ankle power was found with larger ankle power in the saut de chat. A representative plot of net ankle joint power illustrates this result (Fig. 7.12).

Computing joint powers and muscle powers are useful to assess the mechanical demands of a movement on different joints. The timing in power profiles at joints can also provide information about sequential ordering or coordination of a movement.

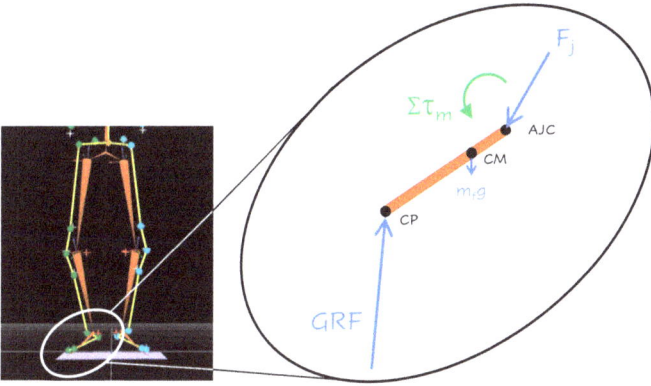

Figure 7.9: Forces and torques acting on a dancer's foot segment during landing from a jump. The ground reaction force (GRF) acts at the center of pressure (CP). The weight of the foot segment ($F_g = m_f g$) acts at the foot's center of mass (CM). The joint force (F_j) is the force on the foot due to the shank (lower leg) segment and acts at the ankle joint center (AJC). The net muscle torque $\Sigma \tau_m$ acts in the plantarflexion direction.

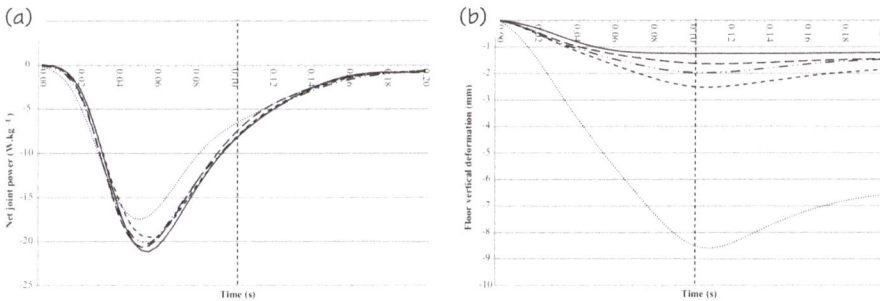

Figure 7.10: (a) Net muscle torque power of the ankle plantarflexors as a function of time during jump landings on five different floor types. (b) Vertical deformation of the floor surface vs. time for the five floor types. Note that the floor with the greatest deformation corresponds to net muscle power with the least negative peak power. Reprinted with permission from [35].

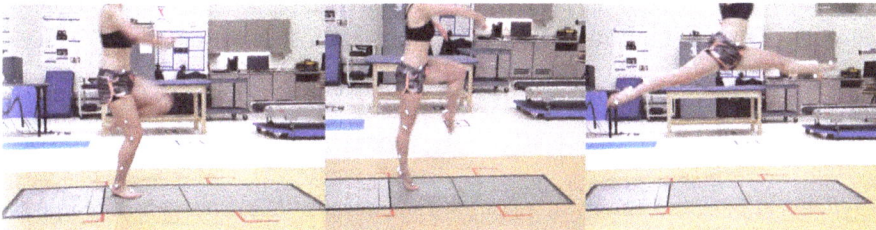

Figure 7.11: Ballet saut de chat is a 2D travelling leap. The ankle plantar flexors, knee and hip extensors of the dancer's right leg generate power during pushoff to propel the dancer's whole body CM into the air for the leap. Reprinted with permission from [77]. Open Access Journal of Sports Medicine 2019 10 191–197' Originally published by and used with permission from Dove Medical Press Ltd.

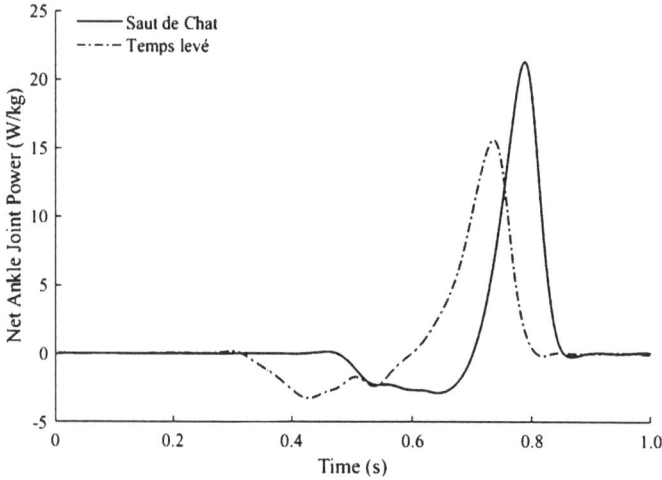

Figure 7.12: Net muscle torque power of the ankle plantarflexors as a function of time during two types of ballet leaps, the saut de chat and temps levé. Reprinted with permission from [77]. Open Access Journal of Sports Medicine 2019 10 191–197' Originally published by and used with permission from Dove Medical Press Ltd.

7.4 Potential energy and mechanical energy

A force is termed conservative if it depends only on position (and not other factors such as velocity or time) and if the work done by the force is path independent (i. e., only depends on the beginning and end points and not the path taken to get from point A to point B). Two examples of conservative forces are the force of gravity and spring force that obeys Hooke's law ($F_s = -kx$). If a force is conservative, we can define a potential energy function associated with that force. Imagine a dancer is lifted by a partner at a constant velocity. The partner must do positive work against the force of gravity to lift the dancer. If a dancer's CM is raised a vertical displacement Δz, the work done by gravity is $W_g = -mg\Delta z$, so the work done by the partner is $W_{partner} = +mg\Delta z$. We can think of this positive work done by the partner as being stored in the dancer–Earth system. We define the change in gravitational potential energy as the negative of the work done by gravity:

$$\Delta U_g = -W_g = mg\Delta z \tag{7.11}$$

Gravitational potential energy increases as the dancer's CM moves vertically upward and decreases as the CM moves vertically downward, regardless of whether the dancer is being lifted by a partner or moving on their own. We define *gravitational potential energy* as

$$U_g = mgz \tag{7.12}$$

where z is measured from an arbitrarily chosen reference position of $z = 0$. For example, zero gravitational potential energy can be chosen at the position of the ground.

There are not exactly any springs acting on a dancer, however, it has been demonstrated that there can be some elastic potential energy storage in the muscle-tendon unit (e. g., in the foot muscles during running [43]). The muscle-tendon unit cannot however be modeled as an idealized Hooke's law spring, as it acts nonlinearly and its spring stiffness depends on how the muscle-tendon is loaded (e. g., more or less rapidly). We therefore will not associate a potential energy function with muscle-tendon elastic potential energy. (The familiar spring potential energy, $U_s = \frac{1}{2}kx^2$ assumes a linear spring.)

From the work-kinetic energy theorem, we can see that if only conservative forces do work on an system (W_c), and work done by non-conservative forces (W_{nc}) is zero, the total mechanical energy, $E = K + U$ is constant or conserved.

$$\Sigma W = \Delta K$$
$$W_c + W_{nc} = \Delta K$$
$$\text{If} \quad W_{nc} = 0, \quad \text{then}$$
$$W_c = \Delta K$$
$$-\Delta U = \Delta K$$
$$\Delta U + \Delta K = 0$$
$$\Delta(U + K) = 0$$

Another way of stating this is that the total mechanical energy is the same initially as it is finally, so

$$\text{If} \quad W_{nc} = 0, \quad \text{then}$$
$$U_i + K_i = U_f + K_f$$

(7.13)

This is a statement of *conservation of mechanical energy*.

If we assume the only force acting on the dancer is gravity (e. g., the dancer is in the air), conservation of mechanical energy applies to the dancer's CM motion. The magnitude of the CM velocity (speed) when a dancer leaves the ground for a 2D leap is $v_1 = \sqrt{\dot{x}_1^2 + \dot{z}_1^2}$. At the peak of the jump, the vertical component of velocity $\dot{z}_2 = 0$, whereas $\dot{x}_2 = \dot{x}_1$ since there is no air resistance. We can then use conservation of energy to solve for the vertical displacement of the dancer's CM during the leap (Δz).

$$U_i + K_i = U_f + K_f$$
$$mgz_1 + \frac{1}{2}m(\dot{x}_1^2 + \dot{z}_1^2) = mgz_2 + \frac{1}{2}m\dot{x}_1^2$$
$$\frac{1}{2}m\dot{z}_1^2 = mg\Delta z$$
$$\Delta z = \frac{\dot{z}_1^2}{2g}$$

This is the same answer we obtained from a kinematics of constant acceleration ($\vec{a} = -g\hat{z}$) approach in Chapter 3.

Conservation of mechanical energy does not apply to the motion of the body segments relative to the CM in addition to CM motion during a leap. The dancer's muscles exert non-conservative forces to increase or decrease their kinetic energy as they rotate around joint centers. A dancer can increase or decrease the total kinetic energy of all of the segments while in the air, and change the overall mechanical energy.

Although total mechanical energy may not be conserved, we can still compute the mechanical energy of individual segments as well as the total mechanical energy of the entire dancer at any time. It can be useful to see how the energy may transfer between segments and to different parts of the body throughout movement. In general, dancing consists of translation and rotation of body segments. For a single segment that we assume is a rigid body (e. g., the dancer's right thigh in Fig. 7.13), it's total kinetic energy is the translational kinetic energy of its CM plus its rotational kinetic energy around the segment CM. The kinetic energy of the i-th segment is

$$K_i = (\text{Segment CM Translational KE}) + (\text{Segment Rotational KE around its CM})$$
$$= \frac{1}{2}m_i v_i^2 + \frac{1}{2}I_i \omega_i^2$$

Figure 7.13: The kinetic energy of an individual body segment (like this dancer's right thigh or left shank) can be found by summing the segment's translational kinetic energy and rotational kinetic energy around the segment's CM.

To find the total kinetic energy of the dancer at any time, we simply sum all of the segments' kinetic energies.

$$K = \sum \left(\frac{1}{2} m_i v_i^2 + \frac{1}{2} I_i \omega_i^2 \right)$$

To find the total gravitational potential energy of the dancer, we could sum the potential energies of all the segments.

$$U_g = \sum m_i g z_i$$

This is equivalent to the CM potential energy, which can be shown by multiplying the above expression by $\frac{M}{M} = 1$, where M is the dancer's total body mass.

$$U_g = \sum m_i g z_i = \sum m_i g z_i \frac{M}{M} = Mg \frac{\sum m_i z_i}{M} = Mg z_{CM}$$

where we have used the definition of CM position (eq. (3.1)).

Finally, the dancer's total mechanical energy is given by eq. (7.14).

$$E = Mg z_{CM} + \sum \left(\frac{1}{2} m_i v_i^2 + \frac{1}{2} I_i \omega_i^2 \right) \tag{7.14}$$

Eq. (7.14) can be used to find the total mechanical energy of the dancer at any point in time during any type of dance movement.

7.5 Metabolic energy

It is common knowledge that we need energy from food to keep us alive. But how does food contain energy and how is this energy extracted to promote life and physical activities such as dance? Ultimately, this is a biochemical process, and while this is not a biochemistry textbook, the following is the basic idea behind aerobic respiration. (Anaerobic processes do not require oxygen as discussed in Chapter 2, but most human respiration is aerobic.) Chemical energy is stored in the bonds of organic molecules like carbohydrates (sugars), proteins, and fats in the food we eat. When those molecules combine with oxygen in a chemical reaction, energy is released. That released energy can then be used as fuel for a whole host of biological processes. For example,

$$\text{Glucose} + \text{Oxygen} \longrightarrow \text{Carbon dioxide} + \text{Water} + \text{Energy}$$
$$C_6 H_{12} O_6 + 6(O_2) \longrightarrow 6(CO_2) + 6(H_2 O) + ATP$$

Metabolism is the process of converting chemical potential energy stored in food into energy that is used in the body. The conventional unit of measurement for energy stored in food is the kilocalorie (kcal), which is the amount of energy required to raise the temperature of 1 kg of water by 1 °C (1 kcal = 4186 J).[1] The energy densities of carbohydrates and protein are 4 kcal/g, and that of fat is 9 kcal/g. Energy densities can be used to determine the overall kcal content of food listed on packaging labels found in grocery stores.

Without delving into complex biochemistry, we will instead separate the energy into four main categories of where the food energy can go once consumed. The first is *basal metabolic energy* or the energy required to sustain basic bodily functions (e. g., breathing, digestion, cell growth, body temperature regulation, brain function, etc.). Second, energy is required to activate skeletal muscle. Recall from Chapter 2 that energy (ATP) is needed to form actin–myosin crossbridges in the sarcomeres within muscle fibers. Third, some of the food energy is not used but instead excreted as waste (approximately 2–9 %, [66]). Finally, energy that is not used or excreted is stored in the body as fat. The term *metabolic energy* characterizes the energy that is used (and not stored or excreted) by the body, so it includes both basal metabolic energy and energy expenditure during muscle activation. Colloquially, metabolic energy is the number of kilocalories that are "burned" by the body whether resting or during exercise.

Dancing often involves both aerobic and anaerobic metabolism by combining slower movements with faster, more powerful movements [84]. Anaerobic metabolism uses sugar (glucose) instead of oxygen as fuel, so the chemical reactions are different than the example previously provided for aerobic metabolism. Many factors contribute to the proportion of aerobic to anaerobic metabolic processes used during dance including the dance style, tempo, particular movements executed, the duration of dance, etc.

7.6 Metabolic rate

Since oxygen drives the chemical reactions converting food into energy used by the body in aerobic metabolism, the rate of oxygen consumption during breathing (VO2) is one indicator of the rate of metabolic energy usage or *metabolic rate*. When the dancer was breathing hard while playing tennis, her lungs were acquiring more oxygen to fuel her muscle's extra energy needs to run, make stops and turns, and swing the racket. To determine energy expenditure during exercise, biomechanists, physiologists, other researchers and clinicians can measure volume of oxygen consumption (and carbon dioxide expulsion) with devices that go over the nose and mouth in a method termed *indirect calorimetry*. Current technology allows for indirect calorimetry devices to be small, portable, and wireless which is ideal for activities such as dance. It has been shown that one liter of oxygen gas consumed corresponds to approximately 4.8 kcal of metabolic energy expenditure.

---

1 One Calorie (spelled with an uppercase "C") is equivalent to one kilocalorie. The Calories on nutrition labels in the U. S. are uppercase Calories. Notably (and regrettably) this is quite confusing, so the use of uppercase Calorie will not be used in this book.

$$1 \text{ Liter } O_2 \approx 4.8 \text{ kcal} \tag{7.15}$$

To date, measuring gas exchanges is the most reliable and widely accepted method to estimate metabolic energy expenditure. Measures of metabolic rate, normalized to body mass are often reported as kcal/kg·hr. Also standard in the literature is to report rate of oxygen consumption in mL/kg·min. Unit conversions and eq. (7.15) can be used to convert between the two quantities. Other devices and methods exist to predict energy expenditure based on accelerometry, heart rate, skin temperature and other measures, although these have been shown to be less reliable than indirect calorimetry, particularly at high intensities of exercise [86].

While resting, our bodies have a baseline rate of energy expenditure, or *resting metabolic rate* (RMR), to continue fueling basic, life-sustaining metabolic processes. Although they are slightly different, basal metabolic energy is often used interchangeably with resting metabolic energy since resting metabolic rate can be measured experimentally. To isolate the *additional* metabolic energy required to activate the muscles and dance (or perform any other physical activity), it is important to know an individual's RMR. A standard way to report the energy expenditure requirements for various activities like dancing or playing tennis is through a quantity called a metabolic equivalent, or MET. A *metabolic equivalent* is defined as the ratio of the total metabolic rate (MR) during an activity to the resting metabolic rate

$$\text{MET} = \frac{\text{MR}}{\text{RMR}} \tag{7.16}$$

Metabolic equivalent is a unitless quantity, however, it is standard to use "MET" as a unit. For example, a brisk walk is approximately 4 METs, which means that the body will be using 4 times as much metabolic energy (per unit time) as it did while resting. Playing a game of singles tennis, like our dancer at the beginning of the chapter, is ~8 METs, requiring 8 times the energy as resting and twice as much metabolic energy as walking for the same amount of time. Light exercise is typically categorized as less than 3 METs, moderate is between 3–6 METs, and vigorous exercise is greater than or equal to 6 METs.

A reference for the MET levels of a variety of activities has been compiled by Ainsworth, et al. (2011) in a "Compendium of Physical Activities" [2]. Many MET values reported in the Compendium were compiled from indirect calorimetry measurements published in the scientific literature. Table 7.3 contains MET values for different types of dance as well as some other activities for comparison. All of the values provided here are experimental measures, either listed in the Compendium or from other dance-specific studies published after 2011. For example, the values for hula dance come from a 2014 study using a portable indirect calorimetry device to measure the RMR and V̇O2 for nineteen elite competitive dancers (ages 18–50 years) performing the Native Hawaiian cultural dance of hula [101]. Dancers performed two variations of hula: one low intensity and one high intensity, based on the tempo and choreographic complexity. The average MET was 5.7 (range 3.17–9.77) for the low-intensity routine and 7.55 (range 4.43–12.0) for high-intensity, with a significant difference in energy expenditure between intensities but not between genders.

Table 7.3: Metabolic energy expenditure by activity. Data are from the Compendium of Physical Activities [2], unless cited otherwise.

MET	Activity
1	sleeping
2.3	cello, sitting
2.5	yoga, Hatha
3.5	Caribbean dance (Abakua, Beguine, Bellair, Bongo, Brukin's, Caribbean Quadrills, Dinki Mini, Gere, Gumbay, Ibo, Jonkonnu, Kumina, Oreisha, Jambu)
4.3	walking, 3.5 mph (level, firm surface, walking for exercise)
4.8	tap dance
5	ballet, modern, or jazz, general, rehearsal or class
5.5	Anishinaabe Jingle Dancing
5.7	hula, low intensity [101]
6.8	ballet, modern, or jazz, performance, vigorous effort
7	soccer, casual
7.1	Polynesian dance, average (Maori haka (5.9), Tongan (6), Samoan sasa (6.6), Fijian (7), Tahitian (8.3), Samoan slap (9.6)) [117]
7.6	hula, high intensity [101]
7.8	general dancing (e. g., disco, folk, Irish step dancing, line dancing, polka)
8	basketball, game
8	tennis, singles
9.8	running (10 min/mile)
9.9	sports dance competition, standard style (women)[*] [78]
10	soccer, competitive
11.3	ballroom dancing, competitive, general
12.6	sports dance competition, standard style (men)[*] [78]

[*]Computed from [78] using RMR = 3.5 ml O_2/kg·min.

The ranges of MET values reported in the hula dance study highlight the fact that the METs given in Table 7.3 are averages and can truly only be used as estimates for the metabolic energy requirements for certain styles of dance. Ranges in measured MET values for a given activity can be large due to individual factors. Not all individuals have the same cardiovascular fitness or ability to supply oxygen to the muscles from the respiratory and circulatory systems. Metabolic rate may also differ based on skill level, with experts in general being more efficient in their movements than beginners (e. g., experts may utilize less coactivation). Individually accurate MET values must be measured experimentally.

In addition, ideally an individual's RMR would be measured experimentally from V̇O2 while the individual is resting. Unfortunately an assumed RMR value of 3.5 $\frac{mL\,O_2}{kg\cdot min}$

has been, and continues to be, widely used in both research and clinical settings. This is in spite of the fact that this value is based on the measured RMR of a single individual (a 40-year-old, 70-kg male) from 1960. Factors such as age, sex, body fat composition, cardiorespiratory fitness, whether a person has a condition such as coronary artery disease, if they are taking medication, can affect a person's RMR [25]. A much more recent study of a heterogenous sample of 642 women and 127 men (18–74 yr, 35–186 kg) reported an average RMR in terms of V̇O2 as $(2.6 \pm 0.4)\,\frac{mL\,O_2}{kg \cdot min}$, which is 26 % less than the often assumed $3.5\,\frac{mL\,O_2}{kg \cdot min}$ [14]. In fact, the "standard" $3.5\,\frac{mL\,O_2}{kg \cdot min}$ is more than 2 standard deviations away from the mean reported in this study.

If an experimental measure of an individual's RMR is not possible, there is a better way to estimate it than simply assuming $3.5\,\frac{mL\,O_2}{kg \cdot min}$ or using the average value reported in [14]. The Harris–Benedict equations are the best available method (for healthy individuals in normal weight range) to theoretically predict a person's resting metabolic energy requirements based on their mass, height, age, and sex [9].

The total daily metabolic energy expenditure (ME) for an individual can be estimated using the *Harris–Benedict equations* [32] (eq. (7.17) for males and eq. (7.18) for females) based on their mass (M) in kilograms, height (H) in centimeters, and age (A) in years.

$$\text{daily ME}_{\text{male}} = 66.47\,\text{kcal} + \left(13.75\,\frac{\text{kcal}}{\text{kg}}\right)M + \left(5.003\,\frac{\text{kcal}}{\text{cm}}\right)H - \left(6.755\,\frac{\text{kcal}}{\text{yr}}\right)A \quad (7.17)$$

$$\text{daily ME}_{\text{female}} = 655.09\,\text{kcal} + \left(9.5634\,\frac{\text{kcal}}{\text{kg}}\right)M + \left(1.8496\,\frac{\text{kcal}}{\text{cm}}\right)H - \left(4.6756\,\frac{\text{kcal}}{\text{yr}}\right)A$$
$$(7.18)$$

Example. Use the Harris–Benedict equations to estimate the basal metabolic energy expenditure per day (in kcal) for three individuals: Person 1 (Male, $M = 74.8\,\text{kg}$, $H = 177.8\,\text{cm}$, $A = 40\,\text{yrs}$), Person 2 (Female, $M = 63.5\,\text{kg}$, $H = 165.1\,\text{cm}$, $A = 26\,\text{yrs}$), and Person 3 (Male, $M = 90\,\text{kg}$, $H = 177.8\,\text{cm}$, $A = 40\,\text{yrs}$). Then report each person's RMR in both kcal/kg·hr and mL O_2/kg·min.

Solution. For Person 1 and 3, we use eq. (7.17), and for Person 2, we use eq. (7.18).

$$\text{daily ME}_1 = 66.47\,\text{kcal}$$
$$+ \left(13.75\,\frac{\text{kcal}}{\text{kg}}\right)(74.8\,\text{kg})$$
$$+ \left(5.003\,\frac{\text{kcal}}{\text{cm}}\right)(177.8\,\text{cm})$$
$$- \left(6.755\,\frac{\text{kcal}}{\text{yr}}\right)(40\,\text{yrs})$$
$$= 1714.5\,\text{kcal}$$
$$\text{daily ME}_2 = 655.09\,\text{kcal}$$

$$+ \left(9.5634 \, \frac{\text{kcal}}{\text{kg}} \right) (63.5 \, \text{kg})$$

$$+ \left(1.8496 \, \frac{\text{kcal}}{\text{cm}} \right) (165.1 \, \text{cm})$$

$$- \left(4.6756 \, \frac{\text{kcal}}{\text{yr}} \right) (26 \, \text{yrs})$$

$$= 1446.2 \, \text{kcal}$$

$$\text{daily ME}_3 = 66.47 \, \text{kcal}$$

$$+ \left(13.75 \, \frac{\text{kcal}}{\text{kg}} \right) (90.0 \, \text{kg})$$

$$+ \left(5.003 \, \frac{\text{kcal}}{\text{cm}} \right) (177.8 \, \text{cm})$$

$$- \left(6.755 \, \frac{\text{kcal}}{\text{yr}} \right) (40 \, \text{yrs})$$

$$= 1923.5 \, \text{kcal}$$

We notice that Person 2 requires the least amount of energy per day for basal metabolic functions. Additionally Person 3 is the same height and age as Person 1 (and both are male), but, with a greater body mass, Person 3 expends more basal metabolic energy per day.

Let's now compute each individual's basal (or resting) metabolic rate (RMR).

$$\text{RMR} = \frac{\text{kcal}}{\text{kg} \cdot \text{hr}}$$

$$\text{RMR}_1 = \frac{1714.5 \, \text{kcal}}{(74.8 \, \text{kg})(24 \, \text{hr})} = 0.955 \, \frac{\text{kcal}}{\text{kg} \cdot \text{hr}}$$

$$\text{RMR}_2 = \frac{1446.2 \, \text{kcal}}{(63.5 \, \text{kg})(24 \, \text{hr})} = 0.949 \, \frac{\text{kcal}}{\text{kg} \cdot \text{hr}}$$

$$\text{RMR}_3 = \frac{1923.5 \, \text{kcal}}{(90.0 \, \text{kg})(24 \, \text{hr})} = 0.891 \, \frac{\text{kcal}}{\text{kg} \cdot \text{hr}}$$

We can convert these values to the alternate units of $\frac{\text{mL O}_2}{\text{kg} \cdot \text{min}}$ using standard unit conversions and eq. (7.15).

$$\frac{1 \, \text{kcal}}{\text{kg} \cdot \text{hr}} \cdot \frac{1000 \, \text{mL O}_2}{4.8 \, \text{kcal}} \cdot \frac{1 \, \text{hr}}{60 \, \text{min}} = 3.47 \, \frac{\text{mL O}_2}{\text{kg} \cdot \text{min}}$$

So, the RMRs of the individuals in these units are

$$\text{RMR}_1 = 3.31 \, \frac{\text{mL O}_2}{\text{kg} \cdot \text{min}}$$

$$\text{RMR}_2 = 3.29 \, \frac{\text{mL O}_2}{\text{kg} \cdot \text{min}}$$

$$\text{RMR}_3 = 3.09 \, \frac{\text{mL O}_2}{\text{kg} \cdot \text{min}}$$

All of the people in this example have RMRs less than the "standard" $3.5 \frac{\text{mL O}_2}{\text{kg·min}}$, but greater than the average value of $2.6 \pm 0.4 \frac{\text{mL O}_2}{\text{kg·min}}$ reported in [14].

From the individual estimates of RMR and average METs for given activities, we can estimate the metabolic energy used (calories burned) by an individual doing an activity such as dancing for a certain amount of time.

Example. Person 1 and Person 2 are ballet dancers who spend 2 hours in a day in rehearsal. (a) Estimate the number of kilocalories each person expends during rehearsal. (b) If each dancer performed in dance sport competition for 15 minutes, what would be their metabolic energy expenditure (in kcal)?

Solution. (a) We use the value reported in Table 7.3, that ballet rehearsal requires 5 METs. We can first find each person's metabolic rate (MR) during rehearsal using eq. (7.16), so MR = 5(RMR)

$$MR_1 = 5(RMR_1) = 5\left(0.955 \, \frac{\text{kcal}}{\text{kg·hr}}\right) = 4.755 \, \frac{\text{kcal}}{\text{kg·hr}}$$

$$MR_2 = 5(RMR_2) = 5\left(0.949 \, \frac{\text{kcal}}{\text{kg·hr}}\right) = 4.745 \, \frac{\text{kcal}}{\text{kg·hr}}$$

Since the metabolic rate gives the number of kilocalories burned per kg per hour of activity, we multiply the rate by the dancer's mass and number of hours of activity.

$$Energy_1 = \left(4.755 \, \frac{\text{kcal}}{\text{kg·hr}}\right)(74.8 \, \text{kg})(2 \, \text{hr}) = 711 \, \text{kcal}$$

$$Energy_2 = \left(4.745 \, \frac{\text{kcal}}{\text{kg·hr}}\right)(63.5 \, \text{kg})(2 \, \text{hr}) = 603 \, \text{kcal}$$

It makes sense that although the metabolic rates are very similar, the dancer with a greater mass would require greater energy to move and dance.

(b) In a dance sport competition, Person 1 (male) has a MET = 12.6 and Person 2 (female) has MET = 9.9 (Tab. 7.3). As in part (a), we first find the metabolic rate for each person.

$$MR_1 = 12.6(RMR_1) = 12.6\left(0.955 \, \frac{\text{kcal}}{\text{kg·hr}}\right) = 12.033 \, \frac{\text{kcal}}{\text{kg·hr}}$$

$$MR_2 = 9.9(RMR_2) = 9.9\left(0.949 \, \frac{\text{kcal}}{\text{kg·hr}}\right) = 9.395 \, \frac{\text{kcal}}{\text{kg·hr}}$$

Then, the total metabolic energy expended by each dancer in 15 minutes (0.25 hours) of dancing is

$$Energy_1 = \left(12.033 \, \frac{\text{kcal}}{\text{kg·hr}}\right)(74.8 \, \text{kg})(0.25 \, \text{hr}) = 225 \, \text{kcal}$$

$$Energy_2 = \left(9.395 \, \frac{\text{kcal}}{\text{kg·hr}}\right)(63.5 \, \text{kg})(0.25 \, \text{hr}) = 149 \, \text{kcal}$$

That is quite a lot of energy for only 15 minutes of dancing, but it is due to the high intensity nature of dance sport.

7.7 Metabolic energy expenditure during dance

The MET values in Table 7.3 show that many forms of dance range from moderate to high intensity physical activity, with the highest intensity being ballroom and standard dance sport (which consists of foxtrot, quickstep, tango, and waltz). We also see that in the partnered dancing of standard dance sport, MET requirements are different for different genders. This is due to the different gendered roles in the dance, with the men lifting their partners, for example. In general dance class MET levels are less than performance, which is seen in other sports as well when comparing practice to competition [84].

Studies have shown that dancers (particularly ballet dancers) tend to have lesser aerobic (cardiorespiratory) fitness than many other athletes, and ballet by itself is not good training for aerobic fitness [50]. Anaerobic fitness is the least studied in dance, but ballet dancers have been found to have more slow-twitch muscle fibers than modern dancers, a form that has higher "anaerobic power outputs" than ballet [50]. Dance training historically has not involved supplemental cardiovascular or weight training for fear that dancers' body aesthetic would be negatively impacted. However, injuries have been associated with low levels of both aerobic fitness and strength, and more recent research has shown that supplemental fitness training in dancers reduces injury risk [84]. This indicates that not only is learning dance technique important, but so is developing physical fitness.

One way to assess aerobic fitness is by measuring the maximum rate at which a person consumes oxygen while exercising, or $\dot{V}O2_{max}$. This is done through indirect calorimetry while a person increases exercise intensity as much as they can until they cannot go anymore (e. g., running on a treadmill while steadily increasing the speed). The greater the amount of oxygen that enters your body through the lungs, the more that is available to fuel the chemical reactions to release energy the muscles can use. Research has found professional contemporary dancers have greater $\dot{V}O2_{max}$ than professional ballet dancers [5].

Fig. 7.14 displays indirect calorimetry data for a ballet dancer during a maximal exertion test until reaching $\dot{V}O2_{max}$. As the intensity of the exercise increases, the dancer's rate of oxygen consumption ($\dot{V}O2$, on the x-axis) increases. On the y-axis is a quantity called the ventilation (VE), which is the rate at which air goes into and out of the lungs (in Liters per minute). The harder a person breathes during exercise, the greater the ventilation. We notice that as intensity increases, ventilation also increases, but that the relationship between VE and $\dot{V}O2$ is nonlinear. At first, the dancer's breathing (VE) only increases slightly with increased intensity, as more oxygen is absorbed by the dancer's lungs (labeled "Below Aerobic Zone" in Fig. 7.14). There is an inflection point, called the first ventilatory threshold (VT1), in which the slope of VE vs $\dot{V}O2$ has increased. In order to keep increasing the rate of oxygen consumption, VE has to increase proportionally more in what is referred to as the aerobic zone. The person's breathing rate will increase more noticeably, and they will have more difficulty talking while exercising. Also in the aerobic zone, the lactate in the blood begins to build as available oxygen begins to run out. Recall that lactate is a byproduct of anaerobic metabolism. Blood tests can be performed to measure lactate levels during exercise (with a finger prick), but ventilatory thresholds can be a non-invasive way of estimating lactate thresholds. The next ventilatory threshold (VT2) occurs when increases in ventilation are no longer accompanied by proportional increases in mechanical power output (e. g., a much larger breathing rate is required for a very small increase in treadmill speed). A spike in lactate concentration also occurs close to VT2 as metabolism shifts from aerobic to anaerobic. This ballet dancer's $\dot{V}O2_{max}$ approaches 45 mL/kg·min. Elite endurance athletes typically have $\dot{V}O2_{max} \sim 65–80$ mL/kg·min.

Fig. 7.14 also shows that a typical ballet class is in the low aerobic zone of intensity, rehearsal and center floor portions of class fall in the moderate aerobic intensity zone, and performance is in the high aerobic to anaerobic zones. Independent of dance style, dance performance has been shown to require greater energy expenditure than class or rehearsal [84]. The dancer playing tennis at the beginning of the chapter may not be very cardiovascularly fit if her only form of regular exercise is a low intensity ballet class. Playing a game of singles tennis requires ~8 METs according to Table 7.3, and the vigourous exercise is likely pushing the dancer past her VT2 threshold.

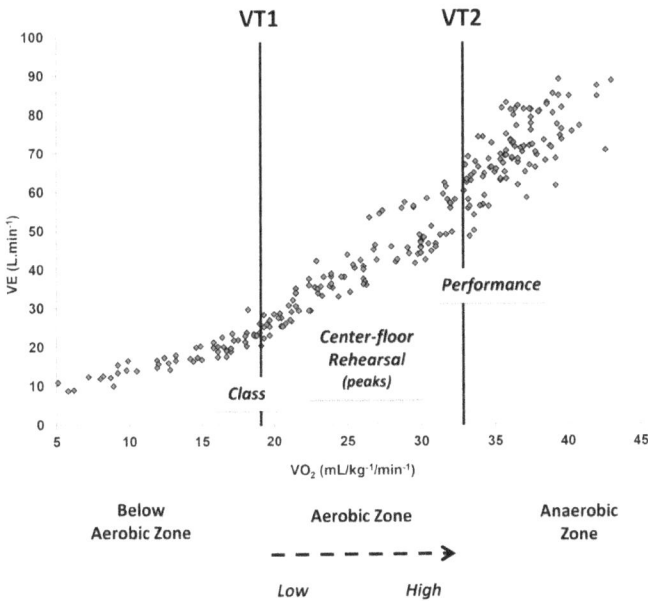

Figure 7.14: Venthillation (VE, volume of gas taken into the lungs per minute) versus rate of oxygen consumption (V̇O2) for a ballet dancer doing a maximum exertion test. Reprinted with permission from [84].

Training just above or below VT2 can benefit a dancer's fitness, so if this does not occur regularly during dance class, the dancer would benefit from supplemental training
(e. g., high-intensity interval training (HIIT)). Dancers additionally need sufficient rest,
both between exercises as well as day to day, so that overtraining does not lead to fatigue
induced injury. Overall health is important for dance health, including getting enough
energy via calories from food to provide energy for movement.

7.8 Energy efficiency

Endurance is needed in dance when classes or rehearsals extend for hours or when a
dancer has a long and taxing piece in a performance. One question a dancer might ask
is how they can move more efficiently while dancing so they are not using metabolic energy unnecessarily. When thinking about efficiency one example that may immediately
come to mind is a car engine. How efficient is a car's engine at converting chemical energy (e. g., from burning gasoline as fuel) into doing useful work (e. g., turning the car's
wheels)? A simple definition of energy efficiency is

$$e = \frac{\text{Mechanical Work Output}}{\text{Energy Input}}$$

Efficiency is always less than 1 because the engine can't create energy from nowhere and
there will always be some energy that goes to sources other than mechanical work (e. g.,
dissipated to friction, thermal energy exhausted). The engine would otherwise violate
the second law of thermodynamics.

In the case of human movement, our muscles are like engines in that they convert
chemical energy into mechanical work done to move our body segments. To quantify
the energy efficiency of our muscles would require measurements of the energy input
(metabolic energy to activate all the skeletal muscles) and the total mechanical work
done by the muscles. Human movement efficiency may be defined as $e = \frac{\Sigma W_m}{(MR-RMR)\Delta t}$.
There are several methods biomechanists have proposed to quantify efficiency in hu-
man movement, but details will not be discussed here.

Quantifying an athlete's efficiency during endurance sports like cycling, distance
running, or cross-country skiing can be a useful performance measurement where the
goal of the sport is to be as efficient as possible so the athlete can maximize center of
mass speed over long distances. While certain styles of dance may require endurance, a
performance goal of dance is not its efficiency. In fact, many movements may be choreographically, and therefore purposefully, inefficient. Instead of diving deeper into how
to quantify efficiency experimentally, a qualitative discussion of factors that can contribute to inefficient movement can help dancers understand how they might move
more efficiently in situations which permit it.

Any situation in which muscles are activated but not producing movement is ener-
getically inefficient. Metabolic energy is needed to activate the muscles, but when dis-

placement of body segments is zero, zero mechanical work is being done. Statically held poses by definition have an efficiency of zero. The more challenging the posture (e. g., when greater muscle force must be exerted to hold loads with large moment arms like an extended leg), the more metabolic energy must be used to still produce zero mechanical work. Even when some body segments are in motion, muscles that act as fixators and prevent motion at other segments contribute to inefficiency. The muscle activation needed to hold segments steady (e. g., of the pelvis and upper body when only the lower half of the body moves) is inefficient in this sense. Tension in the arms is often required for proper technique and aesthetics, even though it is not energy efficient. This is not to say that maintaining proper alignment is inefficient in general. In fact, poor postural alignment requires greater metabolic energy for muscles to be active to support poor posture as opposed to a posture where the body segments are more vertically stacked.

Activation of antagonists during segmental rotation at joints also decreases efficiency of movement. Antagonists do *negative* mechanical work as they exert torque in the opposite direction as rotational velocity. Their activation requires metabolic energy, and, as we learned in Chapter 6, coactivation leads to greater muscle force that is required by agonists (also requiring more metabolic energy). Some activation of antagonists is necessary for slow and controlled movements, so it cannot be eliminated completely. Beginners tend to use more coactivation than experts, which may be one contribution to inefficiency in less experienced dancers. The same kinematics may be achieved with different muscle activation patterns, but more skilled dancers likely use less metabolic energy to perform the same dance-specific task.

Lack of neuromuscular coordination can also lead to inefficiency and is another reason why expert dancers may be more efficient than beginners. Mechanical energy can be more smoothly transferred from one segment to another with sequential muscle firing, but if the muscle activation is not appropriately timed, more energy can be lost between segments. More practice can lead to improved neuromuscular coordination and less coactivation, which in turn can increase a dancer's energy efficiency. This may also be a reason why the dancer was very winded while playing tennis. If she does not regularly play tennis, she likely does not have the same level of neuromuscular coordination when swinging the racket and during tennis-specific movements as someone who is an experienced tennis player. On the other hand, she is likely much more energy efficient when performing her regularly practiced dance movements.

8 Lagrangian mechanics and applications in dance

All of our analyzes of dance movement thus far have remained grounded in Newtonian mechanics, the central principle of which is based on the experimental observation that a system's acceleration is directly proportional to net force and inversely proportional to mass. Biomechanical analyses grounded in Newtonian mechanics are extremely valuable, but Newton's second law is not the only way to approach a problem and explain how and why things move in classical mechanics. This chapter provides a brief introduction to an alternate approach – Lagrangian Mechanics – with applications to a few examples in dance. One approach is not necessarily better than the other, but each has their strengths in different contexts.

Lagrangian Mechanics, like Newtonian mechanics, predicts how things move, but from a different "why." Instead of a net force causing a mass to accelerate, the Lagrangian formulation asserts that the path taken minimizes something called the action. We do not obtain different results depending on whether we analyze the situation with Newtonian or Lagrangian mechanics. The motion of the system is the same; but it can be more or less convenient to analyze a problem from one or the other approach. Newtonian mechanics can feel more comfortable because of its grounding in the familiar and experiential $\Sigma \vec{F} = m\vec{a}$. We have everyday experience with concepts of force, mass, and acceleration, and as humans can feel forces acting on the body. Action, on the other hand, is more mysterious and is not something we directly perceive.

One potential downside of Newton's second law is that its mathematics can become cumbersome, especially in non-Cartesian coordinates. There are occasions when certain symmetries in a problem lead to a better choice in using other coordinate systems (e. g., spherical polar coordinates), but the acceleration in Newton's second law must be rewritten in that new coordinate system, and then it may become difficult to use in practice. An advantage of using Lagrangian mechanics is that unlike Newton's second law, Lagrange's equations are the same for any coordinate system. Another advantage of the Lagrangian approach is that constraint forces are not included in Lagrange's equations (i. e., forces that constrain the particle to a certain path like the force that the rigid rod exerts on the point mass in our simple inverted pendulum model for balance). The true beauty of Lagrangian mechanics will not be revealed with the few examples covered here, nor is the intention of this chapter to derive Lagrange's equation in detail from the calculus of variations. For more details, interested readers are encouraged to consult a textbook such as Taylor's *Classical Mechanics* [97].

This chapter intends to provide readers with enough of a background to see how Lagrangian mechanics can be used in application to a few examples in dance. One example is the by now very familiar inverted pendulum. We have seen this model in more than one context already, so it will provide us with an opportunity to try our new approach on a familiar system. The other is a further extension of Chapter 5 on rigid body rotations. We will model a dancer performing whole body rotations as a symmetric top and use the Lagrangian approach to dive deep into this 3D rigid body rotation. These

https://doi.org/10.1515/9783110642292-008

are not the only potential applications of Lagrangian mechanics to human movement, but are simply meant to demonstrate to readers its use.

8.1 Principle of least action and the Lagrangian

Nature tends to minimize certain functions. Light travels a path that minimizes the time it takes to go from one point to another. If light reflects off of a mirror, the angle of incidence equals the angle of reflection, much like a bouncing ball. But *why* does the light bounce off at an equal angle in the law of reflection? It can be shown that this path, as opposed to the infinite other paths that the light could have taken, is the one that minimizes the time taken to get from the initial to final point. The same is true for why light bends when it travels from one medium to another (e. g., from air to water). The law of refraction (Snell's law, $n_1 \sin \theta_1 = n_2 \sin \theta_2$) can be derived from the fact that, once again, the actual path taken minimizes the light's time of travel.

The central principle of Lagrangian mechanics is that the actual path taken by a particle (or particles) in a time interval t_1 to t_2 minimizes the *action* (S),

$$S = \int_{t_1}^{t_2} \mathcal{L}\, dt \tag{8.1}$$

where $\mathcal{L}$ is the Lagrangian. The *Lagrangian* is defined as

$$\mathcal{L} = K - U \tag{8.2}$$

where K is kinetic energy and U is potential energy. Since the Lagrangian depends on kinetic and potential energies, $\mathcal{L}$ is in general a function of the position(s) and speed(s) of the particle(s) in the system. For example, let's construct the Lagrangian for a single particle of mass m that is free to move (unconstrained) in all three dimensions under the influence of a net force that is conservative (e. g., gravity). The potential energy associated with this force is a function of its 3-D position, $U = U(x, y, z)$. The kinetic energy is

$$K = \frac{1}{2}mv^2 = \frac{1}{2}m\vec{v} \cdot \vec{v} = \frac{1}{2}m(\dot{x}^2 + \dot{y}^2 + \dot{z}^2)$$

So, the Lagrangian is

$$\mathcal{L} = \frac{1}{2}m(\dot{x}^2 + \dot{y}^2 + \dot{z}^2) - U(x, y, z) \tag{8.3}$$

$\mathcal{L}$ is a function of the coordinates x, y, z and their time derivatives $\dot{x}$, $\dot{y}$, $\dot{z}$ (i. e., $\mathcal{L}(x, y, z; \dot{x}, \dot{y}, \dot{z})$). We could have chosen other coordinates (e. g., spherical polar coordinates (r, θ, ϕ)) to describe the position or configuration of the particle, in which case

$\mathcal{L}(r, \theta, \phi; \dot{r}, \dot{\theta}, \dot{\phi})$. Whatever specific coordinate system we choose to use, we would need three *generalized coordinates*, q_1, q_2, q_3, to specify the unconstrained 3D position of the particle. While it will not be proven here, it can be shown that the *Lagrange equations*

$$\frac{\partial \mathcal{L}}{\partial q_i} - \frac{d}{dt}\frac{\partial \mathcal{L}}{\partial \dot{q}_i} = 0 \qquad (8.4)$$

solve for the path taken that minimizes the action.

We will have one Lagrange equation for each generalized coordinate q_i. We can show that Lagrange's equations reduce to Newton's second law using the Lagrangian for unconstrained motion of a point particle (eq. (8.3)). Let's begin with our first generalized coordinate, $q_1 = x$. We find the partial derivative of the Lagrangian with respect to x,

$$\frac{\partial \mathcal{L}}{\partial x} = -\frac{\partial U}{\partial x}$$

and the partial derivative of the Lagrangian with respect to $\dot{x}$

$$\frac{\partial \mathcal{L}}{\partial \dot{x}} = \frac{1}{2}m(2\dot{x}) = m\dot{x}$$

Eq. (8.4) becomes

$$\frac{\partial \mathcal{L}}{\partial x} - \frac{d}{dt}\frac{\partial \mathcal{L}}{\partial \dot{x}} = 0$$
$$-\frac{\partial U}{\partial x} - \frac{d}{dt}m\dot{x} = 0$$
$$-\frac{\partial U}{\partial x} = \frac{d}{dt}m\dot{x}$$

We note that $F_x = -\frac{\partial U}{\partial x}$ is the x-component of force, and $p_x = m\dot{x}$ is the x-component of momentum, so we have

$$F_x = \frac{dp_x}{dt}$$

which is precisely Newton's second law applied to the x-direction. When we apply eq. (8.4) to the other generalized coordinates y and z, we will obtain the y and z components of Newton's second law.

$$\frac{\partial \mathcal{L}}{\partial x} = \frac{d}{dt}\frac{\partial \mathcal{L}}{\partial \dot{x}} \implies F_x = \frac{dp_x}{dt}$$
$$\frac{\partial \mathcal{L}}{\partial y} = \frac{d}{dt}\frac{\partial \mathcal{L}}{\partial \dot{y}} \implies F_y = \frac{dp_y}{dt}$$
$$\frac{\partial \mathcal{L}}{\partial z} = \frac{d}{dt}\frac{\partial \mathcal{L}}{\partial \dot{z}} \implies F_z = \frac{dp_z}{dt}$$

In this simple example, we found that $\frac{\partial \mathcal{L}}{\partial q_i}$ was exactly the i-th component of force on the particle. It is not in general true that $\frac{\partial \mathcal{L}}{\partial q_i}$ must be force (or even have units of force) for any given Lagrangian. $\frac{\partial \mathcal{L}}{\partial q_i}$ may end up being a torque or some other force-like quantity. For this reason, $\frac{\partial \mathcal{L}}{\partial q_i}$ is termed a *generalized force*. Similarly, $\frac{\partial \mathcal{L}}{\partial \dot{q}_i}$ need not be a component of linear momentum, but will be a momentum-like quantity, so is termed a *generalized momentum*.

8.2 Inverted pendulum revisited

Let's now use our Lagrangian approach to reconsider a familiar example – the inverted simple pendulum. In Chapter 4, we constrained the IP to move in a 2D plane to model a dancer toppling in an non-rotating upright position. In Chapter 5, we held the topple angle θ fixed and allowed the IP to rotate around the z-axis, which helped us understand the meaning of products of inertia and the inertia tensor. In this example, we allow the pendulum to move in three dimensions, so it can both topple (changing θ) and rotate around the vertical (changing ϕ) (Fig. 8.1). There is a single constraint in this example, which is that the mass remains at a fixed distance $r = l$ from the origin (meaning that the dancer's CM maintains the same distance from the point of support at the floor). Because of this constraint, we only need two generalized coordinates to specify the configuration of the system at any time: θ and ϕ. Our goal in this example will therefore be to use Lagrange's equations to find and solve equations of motion for θ and ϕ as functions of time. In this way we can find how the dancer both topples and rotates around z given some initial conditions, for example, in a pirouette.

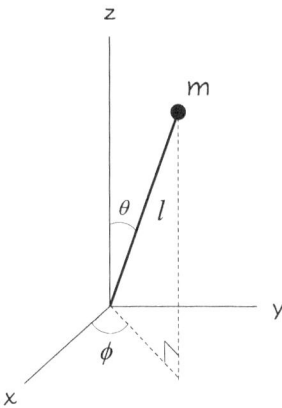

Figure 8.1: Inverted spherical pendulum. Both θ and ϕ can change in this example.

We begin by writing the x, y, z coordinates of the point mass in terms of the generalized coordinates.

$$x = l \sin \theta \cos \phi$$
$$y = l \sin \theta \sin \phi$$
$$z = l \cos \theta$$

Differentiating to find $\dot{x}$, $\dot{y}$, and $\dot{z}$ in terms of θ and ϕ, we have

$$\dot{x} = l(\cos \theta \cos \phi)\dot{\theta} - l(\sin \theta \sin \phi)\dot{\phi}$$
$$\dot{y} = l(\cos \theta \sin \phi)\dot{\theta} + l(\sin \theta \cos \phi)\dot{\phi}$$
$$\dot{z} = -l(\sin \theta)\dot{\theta}$$

The kinetic energy is found from

$$K = \frac{1}{2}m\vec{v} \cdot \vec{v} = \frac{1}{2}m(\dot{x}^2 + \dot{y}^2 + \dot{z}^2)$$

and can be shown to simplify to

$$K = \frac{1}{2}ml^2(\dot{\theta}^2 + \sin^2 \theta \dot{\phi}^2)$$

Notice that this is the sum of rotational kinetic energies around a topple axis and the z-axis.

$$K = \frac{1}{2}ml^2\dot{\theta}^2 + \frac{1}{2}m(l \sin \theta)^2\dot{\phi}^2$$
$$= \frac{1}{2}I_{\text{topple}}\dot{\theta}^2 + \frac{1}{2}I_z\dot{\phi}^2$$

The potential energy is simply $U = mgz = mgl \cos \theta$, so the Lagrangian is

$$\mathcal{L} = K - U = \frac{1}{2}ml^2(\dot{\theta}^2 + \sin^2 \theta \dot{\phi}^2) - mgl \cos \theta$$

The Lagrange equation for the topple angle θ is

$$\frac{\partial \mathcal{L}}{\partial \theta} - \frac{d}{dt}\frac{\partial \mathcal{L}}{\partial \dot{\theta}} = 0$$

$$ml^2 \sin \theta \cos \theta \dot{\phi}^2 + mgl \sin \theta - \frac{d}{dt}(ml^2\dot{\theta}) = 0$$

$$ml^2 \sin \theta \cos \theta \dot{\phi}^2 + mgl \sin \theta - (ml^2\ddot{\theta}) = 0$$

So, the equation of motion for θ is

$$\ddot{\theta} = \sin \theta \cos \theta \dot{\phi}^2 + \frac{g}{l} \sin \theta \qquad (8.5)$$

We notice that if there is no rotation around the z-axis ($\dot{\phi} = 0$), this reduces to the topple equation of motion found in Chapter 4 for the simple inverted pendulum model for balance, $\ddot{\theta} = \frac{g}{l} \sin \theta$. If $\dot{\phi} \neq 0$, it appears our equations of motion for θ and ϕ will be coupled (i. e., the equation of motion for θ has terms with ϕ and/or $\dot{\phi}$ and the equation of motion for ϕ has terms with θ and/or $\dot{\theta}$).

The Lagrange equation for the angle ϕ is

$$\frac{\partial \mathcal{L}}{\partial \phi} - \frac{d}{dt} \frac{\partial \mathcal{L}}{\partial \dot{\phi}} = 0$$

$$0 - \frac{d}{dt}(ml^2 \sin^2 \theta \dot{\phi}) = 0$$

This means that the quantity $ml^2 \sin^2 \theta \dot{\phi}$ does not change with time. While it may not be obvious at first glance, this quantity has precise physical meaning, and is the vertical component of the mass's rotational momentum ($L_z = I_z \dot{\phi}$), which can be verified with $\vec{L} = \vec{r} \times \vec{p}$. This means that L_z is conserved, which makes physical sense because the only torque on the system is due to the force of gravity ($\vec{F}_g = -mg\hat{z}$), and $\vec{\tau}_g = \vec{r} \times \vec{F}_g$ can only ever be in a direction perpendicular to $\vec{F}_g$.

$$L_z = ml^2 \sin^2 \theta \dot{\phi} = \text{constant} \tag{8.6}$$

If the IP begins its motion with a known initial topple angle θ_0 and rate of rotation around the z-axis $\dot{\phi}_0$, we can use the statement of conservation of momentum in eq. (8.6) to determine $\dot{\phi}$ at any other topple angle θ later in the motion.

$$L_z = L_{z,0}$$
$$ml^2(\sin^2 \theta)\dot{\phi} = ml^2(\sin^2 \theta_0)\dot{\phi}_0$$
$$\dot{\phi} = \frac{(\sin^2 \theta_0)\dot{\phi}_0}{\sin^2 \theta}$$

Substituting this expression for $\dot{\phi}$ into eq. (8.5) uncouples the equation of motion for θ, resulting in

$$\ddot{\theta} = (\dot{\phi}_0^2 \sin^4 \theta_0) \frac{\cos \theta}{\sin^3 \theta} + \frac{g}{l} \sin \theta \tag{8.7}$$

This equation of motion for θ is independent of mass, as found for the IP model in Chapter 4. The nonlinear differential equation can be solved numerically given the value of l and initial conditions (θ_0, $\dot{\theta}_0$, and $\dot{\phi}_0$). We also wish to find ϕ as a function of time, so before we solve for θ, let's go back to our Lagrange equation for ϕ and perform the differentiation.

$$\frac{d}{dt}(ml^2 \sin^2 \theta \dot{\phi}) = 0$$
$$ml^2 \frac{d}{dt}(\sin^2 \theta \dot{\phi}) = 0$$

$$\frac{d}{dt}(\sin^2\theta\dot\phi) = 0$$

$$2(\sin\theta\cos\theta)\dot\theta\dot\phi + (\sin^2\theta)\ddot\phi = 0$$

Which results in the following (coupled) EOM for ϕ

$$\ddot\phi = \frac{-2(\cos\theta)\dot\theta\dot\phi}{\sin\theta} \tag{8.8}$$

At this point, we can choose initial conditions and numerically solve eq. (8.7) and eq. (8.8). (Note that there is no need to uncouple the equations when solving them simultaneously with, for example, Mathematica's NDSolve.) As an example we have chosen reasonable values for pendulum length and initial conditions for a dancer in a pirouette ($l = 1\,\mathrm{m}$, $\theta_0 = 10° = 0.1745\,\mathrm{rad}$, $\dot\theta_0 = -170°/\mathrm{s} = -2.97\,\mathrm{rad/s}$ (initial velocity toward balance), $\phi_0 = 0$, and $\dot\phi_0 = 2\,\mathrm{rev/s} = 12.57\,\mathrm{rad/s}$). Plots of θ and ϕ as functions of time are shown in Fig. 8.2.

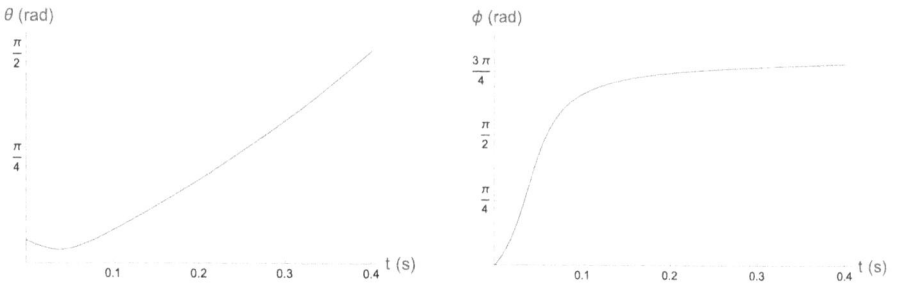

Figure 8.2: Topple angle (θ) and spin angle (ϕ) as functions of time for the spherical IP ($l = 1\,\mathrm{m}$) with initial conditions $\theta_0 = 0.1745\,\mathrm{rad}$, $\dot\theta_0 = -2.97\,\mathrm{rad/s}$, $\phi_0 = 0$, and $\dot\phi_0 = 12.57\,\mathrm{rad/s}$.

We make several observations about how θ and ϕ evolve as the pendulum rotates. First, this would not be a very well executed pirouette, with the dancer toppling to $90°$ after only 0.4 seconds. The dancer additionally would rotate less than half of a revolution ($\phi_{max} \approx \frac{3\pi}{4}$) around the z-axis. Note that in a pirouette, the rotation around the fixed z-axis is actually *not* the same as the rotation of the dancer around their body spin axis (which is ideally the principal axis $\hat e_3$). The dancer's spin axis is affixed to the body, so a dancer can remain spinning around the body axis even if their body's long axis does not rotate around the vertical (which is what $\dot\phi$ measures). In spite of this, we can make a few interesting remarks about the motion of θ and ϕ. We notice that θ first decreases and then increases, which makes sense given that the initial topple velocity was negative, or toward equilibrium. The minimum value of θ (at around $t = 0.05\,\mathrm{s}$) is also where the slope of ϕ vs. t is maximum. This means that the rate of rotation of the pendulum around the z-axis (or precession rate) is greatest at this point in time. This makes physical sense since we discovered that L_z must be conserved. Then as the topple angle increases, $\dot{\phi}$ decreases (slope of ϕ vs. t), once again in line with the conservation of momentum L_z.

8.3 Ignorable coordinates and momentum conservation

We found in the previous example that when $\frac{\partial \mathcal{L}}{\partial \phi} = 0$ this meant that $\frac{\partial \mathcal{L}}{\partial \dot{\phi}}$ remained constant. From the general form of the Lagrange equations in eq. (8.4), we see that

$$\text{If} \quad \frac{\partial \mathcal{L}}{\partial q_i} = 0, \quad \text{then} \quad \frac{d}{dt}\frac{\partial \mathcal{L}}{\partial \dot{q}_i} = 0, \quad \text{and}$$

$$\frac{\partial \mathcal{L}}{\partial \dot{q}_i} = \text{constant}$$

In other words, if the Lagrangian does not have explicit dependence on the generalized coordinate q_i, that means the i-th component of generalized force is zero, and therefore its corresponding generalized momentum ($p_i = \frac{\partial \mathcal{L}}{\partial \dot{q}_i}$) is conserved. If the Lagrangian is independent of a particular generalized coordinate, that generalized coordinate is said to be *ignorable*. This result is particularly useful because if we notice right away that a Lagrangian does not depend on a particular generalized coordinate, we immediately know that its generalized momentum remains constant with time.

8.4 Dancer as a spinning top: Lagrangian approach

In Chapter 5, we found the inertia tensor of a dancer in standard ballet pirouette position. When solving for the principal axes and principal moments of inertia, we found that in this position the two topple moments of inertia were approximately equal with the spin moment of inertia much less ($I_1 = I_2 > I_3$). The dancer in pirouette position shares this property with a symmetric top. With that in mind, we can have a little fun and model the dancer like a spinning top and ask some interesting questions. If we set a top in motion with the size and shape of a dancer, what would be the dynamics of the top's motion? If spun fast enough, would the dancer behave like a "sleeping" top, in which it remains upright for a very long time? Will its spin axis precess or circulate around the vertical like a top? For this analysis, we will treat the entire dancer as a rigid body, in which the shape of the dancer does not change during the pirouette.

The dancer (mass M) is assumed to be symmetric with $I_1 = I_2 = I$ and $I_3 < I$ is the moment of inertia around the dancer's spin principal axis $\hat{e}_3$. The dancer rotates about a fixed point of support at the ground. The topple angle θ and the other two Euler angles, ϕ and ψ describe the position and orientation of the top, and will therefore be our generalized coordinates. ϕ measures the azimuth of the dancer around the vertical and ψ is the rotation angle of the dancer around the spin axis $\hat{e}_3$ (Fig. 8.3). ϕ measures the precession rate of the top's $\hat{e}_3$ axis (longitudinal axis) around z and is not the same as the top's spin rate which is measured by $\dot{\psi}$.

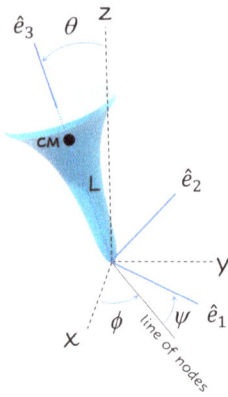

Figure 8.3: Symmetric top with principal topple axes $\hat{e}_1$ and $\hat{e}_2$ and spin principal axis $\hat{e}_3$. The angles θ, ϕ, and ψ are the Euler angles of the top.

Because we assume the point of support is fixed, the top's kinetic energy is purely rotational, $K = \frac{1}{2}\omega^T I\omega$, or

$$K = \frac{1}{2}\begin{pmatrix} \omega_1 & \omega_2 & \omega_3 \end{pmatrix}\begin{pmatrix} I & 0 & 0 \\ 0 & I & 0 \\ 0 & 0 & I_3 \end{pmatrix}\begin{pmatrix} \omega_1 \\ \omega_2 \\ \omega_3 \end{pmatrix} = \frac{1}{2}I(\omega_1^2 + \omega_2^2) + \frac{1}{2}I_3\omega_3^2$$

In terms of Euler angles, $K = \frac{1}{2}I(\dot{\theta}^2 + \dot{\phi}^2 \sin^2\theta) + \frac{1}{2}I_3(\dot{\psi} + \dot{\phi}\cos\theta)^2$. Like the inverted pendulum, the potential energy of the top is given by $U = Mgz$, letting $z = 0$ at the ground. We note $z = L\cos\theta$, where L is the distance of the CM from the point of support. The Lagrangian $\mathcal{L} = K - U$ is therefore

$$\mathcal{L} = \frac{1}{2}I(\dot{\theta}^2 + \dot{\phi}^2 \sin^2\theta) + \frac{1}{2}I_3(\dot{\psi} + \dot{\phi}\cos\theta)^2 - MgL\cos\theta$$

Note that the slowing torque due to friction $\sum \tau_3 = 0$ for this analysis. Also note that ϕ and ψ are ignorable (they do not appear explicitly in $\mathcal{L}$), so their corresponding generalized momenta are constant in time ($p_\phi = \frac{\partial \mathcal{L}}{\partial \dot{\phi}}$ and $p_\psi = \frac{\partial \mathcal{L}}{\partial \dot{\psi}}$).

Assuming friction is negligible, the system is conservative, and another constant in time is the total mechanical energy, $E = K + U$, of the system:

$$E = \frac{1}{2}I(\dot{\theta}^2 + \dot{\phi}^2 \sin^2\theta) + \frac{1}{2}I_3(\dot{\psi} + \dot{\phi}\cos\theta)^2 + MgL\cos\theta$$

Note that this analysis follows closely to that found in Goldstein, Poole, and Safko's *Classical Mechanics* [29]. The following constants of the motion can be defined: $\alpha = \frac{2E - I_3\omega_3^2}{I}$,

$\beta = \frac{2MgL}{I}$, $a = \frac{p_\psi}{I}$, $b = \frac{p_\phi}{I}$. In terms of these constants and defining the variable $u = \cos\theta$, the energy equation becomes

$$\dot{u}^2 = (1 - u^2)(a - \beta u) - (b - au)^2 \tag{8.9}$$

At this point, the qualitative nature of the dancer's motion can be revealed without having to explicitly solve the energy equation. The roots of eq. (8.9) give the turning angles, or where θ changes sign. The numerical values of the constants a, β, a, and b can be found from the dancer's physical characteristics and assuming typical initial conditions for pirouettes. Let's take for our example a male dancer with $M = 82\,\text{kg}$, $H = 1.8\,\text{m}$, $I = 139.9\,\text{kg·m}^2$, and $I_3 = 1.79\,\text{kg·m}^2$. (This is the same dancer whose princpal moments of inertia were found in Section 5.9). For dancers performing pirouettes, once they reach their full pirouette position to begin the turn phase, θ_0 is typically ~5°. We also will assume that the dancers' initial topple speed, $\dot{\theta}_0$, and initial precession rate, $\dot{\phi}_0$, are approximately zero (i. e., the only initial rotational velocity component is around the spin axis), so the total energy, $E = E_0$ is

$$E = E_0 = \frac{1}{2}I_3\omega_3^2 + MgL\cos\theta_0$$

For this value of E,

$$a = \frac{2E - I_3\omega_3^2}{I} = \frac{2MgL\cos\theta_0}{I} = \frac{2(82\,\text{kg})(9.8\,\text{m/s}^2)(1.0\,\text{m})\cos(5°)}{139.9\,\text{kg·m}^2} = 11.44\,\text{s}^{-2}$$

$$\beta = \frac{2MgL}{I} = \frac{2(82\,\text{kg})(9.8\,\text{m/s}^2)(1.0\,\text{m})}{139.9\,\text{kg·m}^2} = 11.49\,\text{s}^{-2}$$

To calculate a and b, generalized momenta are first found:

$$p_\psi = \frac{\partial\mathcal{L}}{\partial\dot{\psi}} = I_3\omega_3$$

$$p_\phi = \frac{\partial\mathcal{L}}{\partial\dot{\phi}} = (I\sin^2\theta_0 + I_3\cos^2\theta_0)\dot{\phi}_0 + I_3\dot{\psi}_0\cos\theta_0$$

$$\simeq I_3\dot{\psi}_0 = I_3\omega_3$$

Since effects such as nutation and precession become more apparent with increasing spin rate, we will give the dancer a generous initial spin rate (that perhaps pushes the boundaries of what is realistic for a dancer). With this in mind, let's let the spin rate $\omega_3 = 5\,\text{rev/s} = 10\pi\,\text{rad/s}$. We then find the constants a and b.

$$a = b = \frac{p_\psi}{I} = \frac{I_3\omega_3}{I} = \frac{(1.79\,\text{kg·m}^2)(10\pi\,\text{rad/s})}{139.9\,\text{kg·m}^2} = 0.402\,\text{s}^{-1}$$

With these values of a, β, a, and b, the roots of eq. (8.9) are $u = 1.00$ and $u = -0.99$. The roots are subject to the physical constraint that $0 < u \le 1$ since the dancer is supported by a horizontal surface. Since the only physical root is $u = \cos\theta = 1.00$, the only turning angle for the dancer is $\theta = 0$, when the dancer's CM crosses the vertical. There is no lower bounding angle, so the dancer will not nutate as he spins.

Let's suspend reality and consider a case where the dancer is instead set spinning with an initial spin rate ten times greater than what we considered previously ($\omega_0 = 50$ rev/s $= 100\pi$ rad/s). (This is, of course, not recommended in reality, since it will lead to very large and unsafe centrifugal force on the dancer.) The values of α and β are unchanged, but a and b are ten times greater than before ($a = b = 4.02\,\text{s}^{-1}$). The new roots of eq. (8.9) are $u = 0.417$ and $u = 0.985$, corresponding to topple angles of $\theta = 65°$ and $10°$ (Fig. 8.4). This means that instead of toppling from an upright position to an angle of $\theta = 90°$ and laying on the ground, the rapidly spinning dancer (top) would topple to $\theta = 65°$, which is now a turning point, so instead of continuing to fall, the dancer would bounce back up with decreasing θ until reaching $10°$ (another turning point). So, as the dancer spins, θ would oscillate between $10°$ and $65°$ while remaining upright. This effect is termed nutation and can be observed in actual spinning tops (but not dancers).

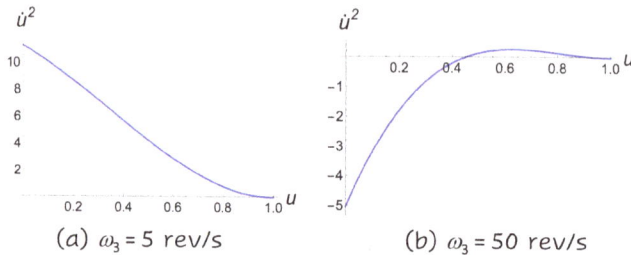

(a) $\omega_3 = 5$ rev/s   (b) $\omega_3 = 50$ rev/s

Figure 8.4: The roots of eq. (8.9) (where the plots above intersect the x-axis) give the turning points of θ for the spinning top.

We can also take this time to determine the initial conditions necessary for steady precession of the dancer-shaped top. For steady precession, the top remains upright at a constant angle $\theta = \theta_0$, so eq. (8.9) must have a double root at $u = u_0$. Mathematically the requirements for a double root are that $f(u) = \dot{u}^2 = 0$ and $\frac{df}{du} = 0$ for $u = u_0$. These requirements lead to a quadratic equation that can be solved for the precession rate:

$$MgL = \dot{\phi}(I_3\omega_3 - I\dot{\phi}\cos\theta_0)$$
$$\longrightarrow (I\cos\theta_0)\dot{\phi}^2 - (I_3\omega_3)\dot{\phi} + MgL = 0$$
$$\dot{\phi} = \frac{I_3\omega_3 \pm \sqrt{(-I_3\omega_3)^2 - 4(I\cos\theta_0)(MgL)}}{2(I\cos\theta_0)}$$

For $\dot{\phi}$ to be real, the quantity under the square root in the above expression must be greater than or equal to zero ($I_3^2\omega_3^2 \geq 4MgLI\cos\theta_0$), or

$$\omega_3^2 \geq \frac{4MgLI\cos\theta_0}{I_3^2} \tag{8.10}$$

For the particular dancer in this example (male, $M = 82\,\text{kg}$, $H = 1.8\,\text{m}$), $\omega_3 \geq 375\,\text{rad/s} = 60\,\text{rev/s}$ for steady precession. An example female dancer ($M = 45\,\text{kg}$ and $H = 1.5\,\text{m}$) would require $\omega_3 \geq 88\,\text{rev/s}$ for steady precession [64]. Typically in a pirouette, dancers rotate at a rate of no greater than a few revolutions per second, much less than the minimum rate required for steady precession based on the inertial properties of a human body in pirouette position. Because of this, complex gyro-dynamics do not need to be factored into the analysis to determine how rapidly a dancer would topple in a pirouette (i. e., to find $\theta(t)$). We are not required to solve eq. (8.9) (by performing elliptic integrals) or beginning with Euler's equations of motion for three-dimensional rotation of a body. The dancer will topple at essentially the same rate as if not spinning [64], with the EOM given by that of the inverted pendulum:

$$\ddot{\theta} = \frac{MgL}{I} \sin\theta \qquad (8.11)$$

For our male example dancer, this results in $\ddot{\theta} = (5.744\,\text{s}^{-2})\sin\theta$. Solving numerically for several example initial conditions, θ vs. time is shown in Fig. 8.5.

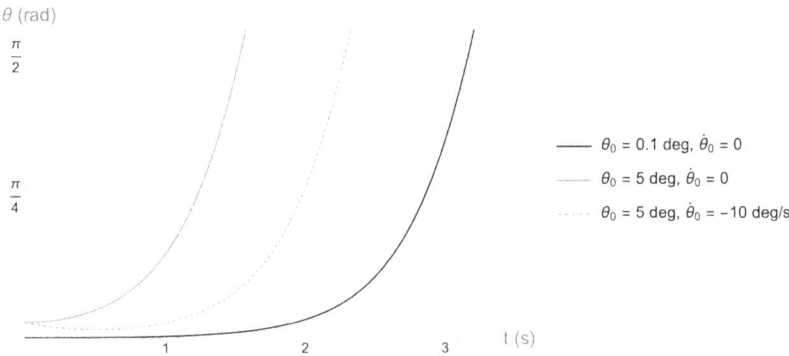

Figure 8.5: Theoretical topple angle as a function of time for a dancer rotating in a pirouette as a rigid body (with no adjustments for balance maintenance). Several examples of initial conditions are given.

Throughout the book, we have created a variety of physical and biomechanical models to represent dancers (or parts of dancers) and analyze their motion. At times the models were knowingly oversimplistic, but could still lead to informative results. The spinning top is one such example of our model allowing us to recognize what is *not* possible for a dancer. It is not realistic for a dancer to behave like a top and remain upright for long periods of time passively spinning. Before performing this analysis we may have incorrectly assumed that if only a dancer spins slightly faster in a pirouette, they could be more like a top making it possible for the turn phase to be a passive aspect of the pirouette. This is very much not the case, however, and a dancer must instead make active adjustments for balance maintenance while rotating.

Bibliography

[1]   ABERGEL, R. E., TUESTA, E., AND JARVIS, D. N. The effects of acute physical fatigue on sauté jump biomechanics in dancers. *Journal of Sports Sciences* 39 (2021), 1021–1029.

[2]   AINSWORTH, B. E., HASKELL, W. L., HERRMANN, S. D., MECKES, N., BASSETT, D. R., TUDOR-LOCKE, C., GREER, J. L., VEZINA, J., WHITT-GLOVER, M. C., AND LEON, A. S. 2011 compendium of physical activities: A second update of codes and MET values. *Medicine and Science in Sports and Exercise* 43 (2011), 1575–1581.

[3]   AMBEGAONKAR, J. P., CASWELL, S. V., WINCHESTER, J. B., SHIMOKOCHI, Y., CORTES, N., AND CASWELL, A. M. Balance comparisons between female dancers and active nondancers. *Research Quarterly for Exercise and Sport* 84 (2013), 24–29.

[4]   AMBEGAONKAR, J. P., CORTES, N., CASWELL, S. V., AMBEGAONKAR, G. P., AND WYON, M. Lower extremity hypermobility, but not core muscle endurance influences balance in female collegiate dancers. *International Journal of Sports Physical Therapy* 11 (2016), 220–229.

[5]   ANGIOI, M., METSIOS, G., KOUTEDAKIS, Y., AND WYON, M. A. Fitness in contemporary dance: A systematic review. *International Journal of Sports Medicine* 30 (2009), 475–484.

[6]   ARMSTRONG, R., BROGDEN, C. M., MILNER, D., NORRIS, D., AND GREIG, M. The influence of fatigue on star excursion balance test performance in dance. *Journal of Dance Medicine & Science* 22 (2018), 142–147.

[7]   ASSLÄNDER, L., AND PETERKA, R. J. Sensory reweighting dynamics in human postural control. *Journal of Neurophysiology* 111 (2014), 1852–1864.

[8]   BASTON, G. Update on proprioception considerations for dance education. *Journal of Dance Medicine & Science* 13 (2009), 35–41.

[9]   BENDAVID, I., LOBO, D. N., BARAZZONI, R., CEDERHOLM, T., COËFFIER, M., DE VAN DER SCHUEREN, M., FONTAINE, E., HIESMAYR, M., LAVIANO, A., PICHARD, C., AND SINGER, P. The centenary of the Harris–Benedict equations: How to assess energy requirements best? Recommendations from the ESPEN expert group. *Clinical Nutrition* 40 (2021), 690–701.

[10]   BHARNUKE, J. K., MULLERPATAN, R. P., AND HILLER, C. Evaluation of standing balance performance in Indian classical dancers. *Journal of Dance Medicine & Science* 24 (2020), 19–23.

[11]   BLENKINSOP, G. M., PAIN, M. T., AND HILEY, M. J. Evaluating feedback time delay during perturbed and unperturbed balance in handstand. *Human Movement Science* 48 (2016), 112–120.

[12]   BOAS, M. L. *Mathematical Methods in the Physical Sciences*, 3rd ed. Wiley, 2005.

[13]   BRIGGS, J., MCCORMACK, M., HAKIM, A. J., AND GRAHAME, R. Injury and joint hypermobility syndrome in ballet dancers – a 5-year follow-up. *Rheumatology* 48 (2009), 1613–1614.

[14]   BYRNE, N. M., HILLS, A. P., HUNTER, G. R., WEINSIER, R. L., SCHUTZ, Y., AND BYRNE, N. M. Metabolic equivalent: One size does not fit all. *Journal of Applied Physiology* 99 (2005), 1112–1119.

[15]   CARDINAL, M. K., ROGERS, K. A., AND CARDINAL, B. J. Inclusion of dancer wellness education programs in U. S. colleges and universities: A 20-year update. *Journal of Dance Medicine & Science* 24 (2020), 73–87.

[16]   CHANDLER, R., CLAUSER, C., MCCONVILLE, J., REYNOLDS, H., AND YOUNG, J. Investigation of inertial properties of the human body, Air Force Aerospace Medical Research Lab Wright-Patterson AFB OH, 1975.

[17]   CHRISTENSEN, S. K., JOHNSON, A. W., WAGONER, N. V., COREY, T. E., MCCLUNG, M. S., AND HUNTER, I. Characteristics of eight Irish dance landings: Considerations for training and overuse injury prevention. *Journal of Dance Medicine & Science* 25 (2021), 30–37.

[18]   CROTTS, D., THOMPSON, B., NAHOM, M., RYAN, S., AND NEWTON, R. A. Balance abilities of professional dancers on select balance tests. *Journal of Orthopaedic and Sports Physical Therapy* 23 (1996), 12–17.

[19]   DAY, H., KOUTEDAKIS, Y., AND WYON, M. A. Hypermobility and dance: A review. *International Journal of Sports Medicine* 32 (2011), 485–489.

[20]   DE LEVA, P. Adjustments to Zatsiorsky–Seluyanov's segment inertia parameters. *Journal of Biomechanics* 29 (1996), 1223–1230.

[21]   DE MELLO, M. C., DE SÁ FERREIRA, A., AND FELICIO, L. R. Postural control during different unipodal positions in professional ballet dancers. *Journal of Dance Medicine & Science* 21 (2017), 151–155.

https://doi.org/10.1515/9783110642292-009

[22] DOSTAL, W. F., AND ANDREWS, J. G. A three-dimensional biomechanical model of hip musculature. *Journal of Biomechanics* 14 (1981), 803–812.

[23] FIETZER, A. L., CHANG, Y. J., AND KULIG, K. Dancers with patellar tendinopathy exhibit higher vertical and braking ground reaction forces during landing. *Journal of Sports Sciences* 30 (2012), 1157–1163.

[24] FOLEY, E. C., AND BIRD, H. A. Hypermobility in dance: Asset, not liability. *Clinical Rheumatology* 32 (2013), 455–461.

[25] FRANKLIN, B. A., BRINKS, J., BERRA, K., LAVIE, C. J., GORDON, N. F., AND SPERLING, L. S. Using metabolic equivalents in clinical practice. *American Journal of Cardiology* 121 (2018), 382–387.

[26] FRIEDERICH, J. A., AND BRAND, R. A. Muscle fiber architecture in the human lower limb. *Journal of Biomechanics* 23 (1990), 91–95.

[27] FUGLEVAND, A. J., WINTER, D. A., AND PATLA, A. E. Models of recruitment and rate coding organization in motor-unit pools. *Journal of Neurophysiology* 70 (1993), 2470–2488.

[28] GAUTIER, G., MARIN, L., LEROY, D., AND THOUVARECQ, R. Dynamics of expertise level: Coordination in handstand. *Human Movement Science* 28 (2009), 129–140.

[29] GOLDSTEIN, H., POOLE, C. P., AND SAFKO, J. L. *Classical Mechanics*, 3rd ed. Addison Wesley, San Francisco, 2002.

[30] HABER, C., AND SCHÄRLI, A. Defining spotting in dance: A Delphi method study evaluating expert opinions. *Frontiers in Psychology* 12 (2021), 540396.

[31] HACKNEY, J., BRUMMEL, S., BECKER, D., SELBO, A., KOONS, S., AND STEWART, M. Effect of sprung (suspended) floor on lower extremity stiffness during a force-returning ballet jump. *Medical Problems of Performing Artists* 26 (2011), 195–199.

[32] HARRIS, J. A., AND BENEDICT, F. G. A biometric study of human basal metabolism. *Proceedings of the National Academy of Sciences* 4 (1918), 370–373.

[33] HINCAPIÉ, C. A., MORTON, E. J., AND CASSIDY, J. D. Musculoskeletal injuries and pain in dancers: A systematic review. *Archives of Physical Medicine and Rehabilitation* 89 (2008), 1819–1829.

[34] HOF, A., GAZENDAM, M., AND SINKE, W. The condition for dynamic stability. *Journal of Biomechanics* 38 (2005), 1–8.

[35] HOPPER, L. S., ALDERSON, J. A., ELLIOTT, B. C., AND ACKLAND, T. R. Dance floor force reduction influences ankle loads in dancers during drop landings. *Journal of Science and Medicine in Sport* 18 (2015), 480–485.

[36] HOPPER, L. S., ALLEN, N., WYON, M., ALDERSON, J. A., ELLIOTT, B. C., AND ACKLAND, T. R. Dance floor mechanical properties and dancer injuries in a touring professional ballet company. *Journal of Science and Medicine in Sport* 17 (2014), 29–33.

[37] HOPPER, D. M., GRISBROOK, T. L., NEWNHAM, P. J., AND EDWARDS, D. J. The effects of vestibular stimulation and fatigue on postural control in classical ballet dancers. *Journal of Dance Medicine & Science* 18 (2014), 67–73.

[38] HUTT, K., AND REDDING, E. The effect of an eyes-closed dance-specific training program on dynamic balance in elite pre-professional ballet dancers: A randomized controlled pilot study. *Journal of Dance Medicine & Science* 18 (2014), 3–11.

[39] IVANENKO, Y., AND GURFINKEL, V. S. Human postural control. *Frontiers in Neuroscience* 12 (2018), 1–9.

[40] KADEL, N. J., TEITZ, C. C., AND KRONMAL, R. A. Stress fractures in ballet dancers. *The American Journal of Sports Medicine* 20 (1992), 445–449.

[41] KAUFMANN, J. E., NELISSEN, R. G., EXNER-GRAVE, E., AND GADEMAN, M. G. Does forced or compensated turnout lead to musculoskeletal injuries in dancers? A systematic review on the complexity of causes. *Journal of Biomechanics* 114 (2021), 110084.

[42] KAYA, M., LEONARD, T., AND HERZOG, W. Coordination of medial gastrocnemius and soleus forces during cat locomotion. *Journal of Experimental Biology* 206 (2003), 3645–3655.

[43] KELLY, L. A., FARRIS, D. J., CRESSWELL, A. G., AND LICHTWARK, G. A. Intrinsic foot muscles contribute to elastic energy storage and return in the human foot. *Journal of Applied Physiology* 126 (2019), 231–238.

[44]   KERWIN, D. G., AND TREWARTHA, G. Strategies for maintaining a handstand in the anterior-posterior direction. *Medicine and Science in Sports and Exercise* 33 (2001), 1182–1188.

[45]   KILBY, M. C., MOLENAAR, P. C. M., AND NEWELL, K. M. Models of postural control: Shared variance in joint and COM motions. *PLoS ONE* 10 (2015), 1–20.

[46]   KILBY, M. C., AND NEWELL, K. M. Intra- and inter-foot coordination in quiet standing: Footwear and posture effects. *Gait & Posture* 35 (2012), 511–516.

[47]   KIM, J., WILSON, M. A., SINGHAL, K., GAMBLIN, S., SUH, C.-Y., AND KWON, Y.-H. Generation of vertical angular momentum in single, double, and triple-turn *pirouette en dehors* in ballet. *Sports Biomechanics* 13 (2014), 215–229.

[48]   KLOSTERMANN, A., SCHÄRLI, A., KUNZ, S., WEBER, M., AND HOSSNER, E. J. Learn to turn: Does spotting foster skill acquisition in pirouettes? *Research Quarterly for Exercise and Sport* 93 (2022), 153–161.

[49]   KOCHANOWICZ, A., NIESPODZIŃSKI, B., MARINA, M., MIESZKOWSKI, J., BISKUP, L., AND KOCHANOWICZ, K. Relationship between postural control and muscle activity during a handstand in young and adult gymnasts. *Human Movement Science* 58 (2018), 195–204.

[50]   KOUTEDAKIS, Y., AND JAMURTAS, A. The dancer as a performing athlete: Physiological considerations. *Sports Medicine* 34 (2004), 651–661.

[51]   KRASNOW, D., AMBEGAONKAR, J. P., WILMERDING, M. V., STECYK, S., KOUTEDAKIS, Y., AND WYON, M. Electromyographic comparison of grand battement devant at the barre, in the center, and traveling. *Medical Problems of Performing Artists* 27 (2012), 143–155.

[52]   KRITYAKIARANA, W., AND JONGKAMONWIWAT, N. Comparison of balance performance between Thai classical dancers and non-dancers. *Journal of Dance Medicine & Science* 20 (2016), 72–78.

[53]   KSHTRIYA, S., BARNSTAPLE, R., RABINOVICH, D. B., AND DESOUZA, J. F. Dance and aging: A critical review of findings in neuroscience. *American Journal of Dance Therapy* 37 (2015), 81–112.

[54]   KULIG, K., FIETZER, A. L., AND POPOVICH, J. M. Ground reaction forces and knee mechanics in the weight acceptance phase of a dance leap take-off and landing. *Journal of Sports Sciences* 29 (2011), 125–131.

[55]   LAWS, K. L. An analysis of turns in dance. *Dance Research Journal* 11 (1978–1979), 12–19.

[56]   LAWS, K. Momentum transfer in dance movement. *Medical Problems of Performing Artists* 13 (1998), 136–145.

[57]   LAWS, K. *Physics and the Art of Dance: Understanding Movement*, 2nd ed. Oxford University Press, 2008.

[58]   LAWS, K., AND LEE, K. The grand jete: A physical analysis. *Kinesiology for Dance* 11 (1989), 12–13.

[59]   LEE, H.-H., LIN, C.-W., WU, H.-W., WU, T.-C., AND LIN, C.-F. Changes in biomechanics and muscle activation in injured ballet dancers during a jump-land task with turnout (sissonne fermée). *Journal of Sports Sciences* 30 (2012), 689–697.

[60]   LEPELLEY, M. C., THULLIER, F., KORAL, J., AND LESTIENNE, F. G. Muscle coordination in complex movements during jeté in skilled ballet dancers. *Experimental Brain Research* 175 (2006), 321–331.

[61]   LIEDERBACH, M., DILGEN, F. E., AND ROSE, D. J. Incidence of anterior cruciate ligament injuries among elite ballet and modern dancers: A 5-year prospective study. *American Journal of Sports Medicine* 36 (2008), 1779–1788.

[62]   LIEDERBACH, M., KREMENIC, I. J., ORISHIMO, K. F., PAPPAS, E., AND HAGINS, M. Comparison of landing biomechanics between male and female dancers and athletes, part 2: Influence of fatigue and implications for anterior cruciate ligament injury. *American Journal of Sports Medicine* 42 (2014), 1089–1095.

[63]   LOTT, M. B. Translating the base of support: A mechanism for balance maintenance during rotations in dance. *Journal of Dance Medicine & Science* 23 (2019), 17–25.

[64]   LOTT, M. B., AND LAWS, K. L. The physics of toppling and regaining balance during a pirouette. *Journal of Dance Medicine & Science* 16 (2012), 167–174.

[65]   LOTT, M. B., AND XU, G. Joint angle coordination strategies during whole body rotations on a single lower-limb support: An investigation through ballet pirouettes. *Journal of Applied Biomechanics* 36 (2020), 103–112.

[66] LUND, J., GERHART-HINES, Z., AND CLEMMENSEN, C. Role of energy excretion in human body weight regulation. *Trends in Endocrinology and Metabolism* 31 (2020), 705–708.

[67] MARULLI, T. A., HARMON-MATTHEWS, L. E., DAVIS-COEN, J. H., WILLIGENBURG, N. W., AND HEWETT, T. E. Eyes-closed single-limb balance is not related to hypermobility status in dancers. *Journal of Dance Medicine & Science* 21 (2017), 70–75.

[68] MAYERS, L., BRONNER, S., AGRAHARASAMAKULAM, S., AND OJOFEITIMI, S. Lower extremity kinetics in tap dance. *Journal of Dance Medicine & Science* 14 (2010), 3–10.

[69] MCKAY, J. L., TING, L. H., AND HACKNEY, M. E. Balance, body motion, and muscle activity after high-volume short-term dance-based rehabilitation in persons with Parkinson disease: A pilot study. *Journal of Neurologic Physical Therapy* 40 (2016), 257–268.

[70] MCPHERSON, A. M., SCHRADER, J. W., AND DOCHERTY, C. L. Ground reaction forces in ballet: Differences resulting from footwear and jump conditions. *Journal of Dance Medicine & Science* 23 (2019), 34–39.

[71] MILLER, H. N., RICE, P. E., FELPEL, Z. J., STIRLING, A. M., BENGTSON, E. N., AND NEEDLE, A. R. Influence of mirror feedback and ankle joint laxity on dynamic balance in trained ballet dancers. *Journal of Dance Medicine & Science* 22 (2018), 184–191.

[72] MORRIN, N., AND REDDING, E. Acute effects of warm-up stretch protocols on balance, vertical jump height, and range of motion in dancers. *Journal of Dance Medicine & Science* 17 (2013), 34–40.

[73] NEUMANN, D. A. Kinesiology of the hip: A focus on muscular actions. *The Journal of Orthopaedic and Sports Physical Therapy* 40 (2010), 82–94.

[74] PAI, Y.-C., AND PATTON, J. Center of mass velocity-position predictions for balance control. *Journal of Biomechanics* 30 (1997), 347–354.

[75] PERALA, H. D., WILSON, M. A., AND DAI, B. The effect of footwear on free moments during a rotational movement in country swing dance. *Journal of Dance Medicine & Science* 22 (2018), 84–90.

[76] PÉREZ, R. M., SOLANA, R. S., MURILLO, D. B., AND HERNÁNDEZ, F. J. M. Visual availability, balance performance and movement complexity in dancers. *Gait & Posture* 40 (2014), 556–560.

[77] PERRY, S. K., BUDDHADEV, H. H., BRILLA, L. R., AND SUPRAK, D. N. Mechanical demands at the ankle joint during saut de chat and temps levé jumps in classically trained ballet dancers. *Open Access Journal of Sports Medicine* 10 (2019), 191–197.

[78] PILCH, W., TOTA, L., POKORA, I., GŁOWA, M., PIOTROWSKA, A., CHLIPALSKA, O., ZUZIAK, R., AND CZERWIŃSKA, O. Energy expenditure and lactate concentration in sports dancers in a simulated final round of the standard style competition. *Human Movement* 18 (2017), 62–67.

[79] PINTER, I. J., SWIGCHEM, R. V., VAN SOEST, A. J. K., AND ROZENDAAL, L. A. The dynamics of postural sway cannot be captured using a one-segment inverted pendulum model: A PCA on segment rotations during unperturbed stance. *Journal of Neurophysiology* 100 (2008), 3197–3208.

[80] PRILUTSKY, B. I., AND ZATSIORSKY, V. M. Optimization-based models of muscle coordination. *Exercise and Sport Sciences Reviews* 30 (2002), 32–38.

[81] RAMSHORST, C. V., AND CHOI, W. J. Characteristics of contact force and muscle activation during choreographed falls with 2 common landing techniques in contemporary dance. *Journal of Applied Biomechanics* 35 (2019), 256–262.

[82] REIN, S., FABIAN, T., ZWIPP, H., RAMMELT, S., AND WEINDEL, S. Postural control and functional ankle stability in professional and amateur dancers. *Clinical Neurophysiology* 122 (2011), 1602–1610.

[83] ROBERTSON, D. G. E., CALDWELL, G. E., HAMILL, J., KAMEN, G., AND WHITTLESEY, S. *Research Methods in Biomechanics*, 2nd ed. Human Kinetics, 2014.

[84] RODRIGUES-KRAUSE, J., KRAUSE, M., AND REISCHAK-OLIVEIRA, Á. Cardiorespiratory considerations in dance: From classes to performances. *Journal of Dance Medicine & Science* 19 (2015), 91–102.

[85] RUSSELL, J. A., AND MUELLER, I. F. Force attenuation properties of padded dance support socks. *Journal of Dance Medicine & Science* 26 (2022), 106–113.

[86] SANTOS-LOZANO, A., HERNÁNDEZ-VICENTE, A., PÉREZ-ISAAC, R., SANTÍN-MEDEIROS, F., CRISTI-MONTERO, C., CASAJÚS, J. A., AND GARATACHEA, N. Is the sensewear armband accurate enough to quantify and estimate energy expenditure in healthy adults? *Annals of Translational Medicine* 5 (2017), 97.

[87] SAWERS, A., ALLEN, J. L., AND TING, L. H. Long-term training modifies the modular structure and organization of walking balance control. *Journal of Neurophysiology* 114 (2015), 3359–3373.

[88] SCHEIDLER, A. M., KINNETT-HOPKINS, D., LEARMONTH, Y. C., MOTL, R., AND LÓPEZ-ORTIZ, C. Targeted ballet program mitigates ataxia and improves balance in females with mild-to-moderate multiple sclerosis. *PLoS ONE* 13 (2018), e0205382.

[89] SCHMIT, J. M., REGIS, D. I., AND RILEY, M. A. Dynamic patterns of postural sway in ballet dancers and track athletes. *Experimental Brain Research* 163 (2005), 370–378.

[90] SETH, A., HICKS, J. L., UCHIDA, T. K., HABIB, A., DEMBIA, C. L., DUNNE, J. J., ONG, C. F., DEMERS, M. S., RAJAGOPAL, A., MILLARD, M., HAMNER, S. R., ARNOLD, E. M., YONG, J. R., LAKSHMIKANTH, S. K., SHERMAN, M. A., KU, J. P., AND DELP, S. L. Opensim: Simulating musculoskeletal dynamics and neuromuscular control to study human and animal movement. *PLoS Computational Biology* 14 (2018), e1006223.

[91] SHIPPEN, J. M., AND MAY, B. Calculation of muscle loading and joint contact forces during the rock step in Irish dance. *Journal of Dance Medicine & Science* 14 (2010), 11–18.

[92] SIMMONS, R. W. Neuromuscular responses of trained ballet dancers to postural perturbations. *International Journal of Neuroscience* 115 (2005), 1193–1203.

[93] SIMMONS, R. W. Sensory organization determinants of postural stability in trained ballet dancers. *International Journal of Neuroscience* 115 (2005), 87–97.

[94] SIMPSON, K. J., AND KANTER, L. Jump distance of dance landings influencing internal joint forces: I. Axial forces. *Medicine and Science in Sports and Exercise* 29 (1997), 916–927.

[95] STEINBERG, N., HERSHKOVITZ, I., ZEEV, A., ROTHSCHILD, B., AND SIEV-NER, I. Joint hypermobility and joint range of motion in young dancers. *Journal of Clinical Rheumatology* 22 (2016), 171–178.

[96] SUGANO, A., AND LAWS, K. Physical analysis as a foundation for pirouette training. *Medical Problems of Performing Artists* 17 (2002), 29–32.

[97] TAYLOR, J. R. *Classical Mechanics*. University Science Books, Sausalito, CA, 2005.

[98] TING, L. H., AND MCKAY, J. L. Neuromechanics of muscle synergies for posture and movement. *Current Opinion in Neurobiology* 17 (2007), 622–628.

[99] TORRES-OVIEDO, G., AND TING, L. H. Muscle synergies characterizing human postural responses. *Journal of Neurophysiology* 98 (2007), 2144–2156.

[100] TREPMAN, E., GELLMAN, R., MICHELI, L., AND DE LUCA, C. Electromyographic analysis of grand-plié in ballet and modern dancers. *Medicine and Science in Sports and Exercise* 30 (1998), 1708–1720.

[101] USAGAWA, T., LOOK, M., DE SILVA, M., STICKLEY, C., KAHOLOKULA, J. K., SETO, T., AND MAU, M. Metabolic equivalent determination in the cultural dance of hula. *International Journal of Sports Medicine* 35 (2014), 399–402.

[102] VALENTINO, R., SAVASTANO, S., TOMMASELLI, A. P., D'AMORE, G., DORATO, M., AND LOMBARDI, G. The influence of intense ballet training on trabecular bone mass, hormone status, and gonadotropin structure in young women. *The Journal of Clinical Endocrinology and Metabolism* 86 (2001), 4674–4678.

[103] VARGAS-MACÍAS, A., BAENA-CHICÓN, I., GORWA, J., MICHNIK, R. A., NOWAKOWSKA-LIPIEC, K., GÓMEZ-LOZANO, S., AND FORCZEK-KARKOSZ, W. Biomechanical effects of flamenco footwork. *Journal of Human Kinetics* 80 (2021), 19–27.

[104] WEIGHART, H., MORROW, N., DIPASQUALE, S., AND IVES, S. J. Examining neuromuscular control of the vastus medialis oblique and vastus lateralis muscles during fundamental dance movements. *Journal of Dance Medicine & Science* 24 (2020), 153–160.

[105] WIELANDT, T., VAN DEN WYNGAERT, T., UIJTTEWAAL, J. R., HUYGHE, I., MAES, M., AND STASSIJNS, G. Bone mineral density in adolescent elite ballet dancers. *The Journal of Sports Medicine and Physical Fitness* 59 (2019), 1564–1570.

[106] WILD, C. Y., GREALISH, A., AND HOPPER, D. Lower limb and trunk biomechanics after fatigue in competitive female Irish dancers. *Journal of Athletic Training* 52 (2017), 643–648.

[107] WILSON, M., DAI, B., ZHU, Q., AND HUMPHREY, N. Trunk muscle activation and estimating spinal compressive force in rope and harness vertical dance. *Journal of Dance Medicine & Science* 19 (2015), 163–172.

[108] WINTER, D. A. Human balance and posture control during standing and walking. *Gait & Posture* 3 (1995), 193–214.

[109] WU, G., SIEGLER, S., ALLARD, P., KIRTLEY, C., LEARDINI, A., ROSENBAUM, D., WHITTLE, M., D'LIMA, D. D., CRISTOFOLINI, L., WITTE, H., SCHMID, O., AND STOKES, I. ISB recommendation on definitions of joint coordinate system of various joints for the reporting of human joint motion – Part I: Ankle, hip, and spine. *Journal of Biomechanics* 35 (2002), 543–548.

[110] WU, G., VAN DER HELM, F. C., VEEGER, H. E., MAKHSOUS, M., VAN ROY, P., ANGLIN, C., NAGELS, J., KARDUNA, A. R., MCQUADE, K., WANG, X., WERNER, F. W., AND BUCHHOLZ, B. ISB recommendation on definitions of joint coordinate systems of various joints for the reporting of human joint motion – Part II: Shoulder, elbow, wrist and hand. *Journal of Biomechanics* 38 (2005), 981–992.

[111] YOUNG, N., FORMICA, C., SZMUKLER, G., AND SEEMAN, E. Bone density at weight-bearing and nonweight-bearing sites in ballet dancers: The effects of exercise, hypogonadism, and body weight. *The Journal of Clinical Endocrinology and Metabolism* 78 (1994), 449–454.

[112] ZAFERIOU, A. M., WILCOX, R. R., AND MCNITT-GRAY, J. L. Modification of impulse generation during piqué turns with increased rotational demands. *Human Movement Science* 47 (2016), 220–230.

[113] ZAFERIOU, A. M., WILCOX, R. R., AND MCNITT-GRAY, J. L. Modification of impulse generation during pirouette turns with increased rotational demands. *Journal of Applied Biomechanics* 32 (2016), 425–432.

[114] ZAFERIOU, A. M., WILCOX, R. R., AND MCNITT-GRAY, J. L. Whole-body balance regulation during the turn phase of piqué and pirouette turns with varied rotational demands. *Medical Problems of Performing Artists* 31 (2016), 96–103.

[115] ZAFERIOU, A. M., FLASHNER, H., WILCOX, R. R., AND MCNITT-GRAY, J. L. Lower extremity control during turns initiated with and without hip external rotation. *Journal of Biomechanics* 52 (2017), 130–139.

[116] ZATSIORSKY, V., SELUYANOV, V., AND CHUGUNOVA, L. Methods of determining mass-inertial characteristics of human body segments. In: Chernyi, G. G., Regirer, S. A. (Eds.) *Contemporary Problems of Biomechanics*. CRC Press, Massachusetts (1990), pp. 272–291.

[117] ZHU, W., LANKFORD, D. E., REECE, J. D., AND HEIL, D. P. Characterizing the aerobic and anaerobic energy costs of Polynesian dances. *International Journal of Exercise Science* 11 (2018), 1156–1172.

Index

https://doi.org/10.1515/9783110642292-010

www.ingramcontent.com/pod-product-compliance
Lightning Source LLC
Chambersburg PA
CBHW061411210326
41598CB00035B/6179